Lecture Notes in Chemistry

Edited by G. Berthier M. J. S. Dewar H. Fischer
K. Fukui G. G. Hall H. Hartmann H. H. Jaffé J. Jortner
W. Kutzelnigg K. Ruedenberg E. Scrocco

35

Wavefunctions
and Mechanisms from
Electron Scattering Processes

Edited by F. A. Gianturco and G. Stefani

Springer-Verlag
Berlin Heidelberg New York Tokyo 1984

Editors

F. A. Gianturco
Università di Roma, Dipartimento di Chimica
Città Universitaria, I-00100 Roma AD

G. Stefani
Istituto M.A.I., C.N.R., Area della Ricerca di Roma
C.P. 10, I-00016 Monterotondo (Roma)

7198 7320

se(l...
chim

ISBN 3-540-13347-X Springer-Verlag Berlin Heidelberg New York Tokyo
ISBN 0-387-13347-X Springer-Verlag New York Heidelberg Berlin Tokyo

Library of Congress Cataloging in Publication Data. Main entry under title: Wavefunctions
and mechanisms from electron scattering processes. (Lecture notes in chemistry; 35) Includes
index. 1. Electrons–Scattering. 2. Electron-molecule scattering. 3. Wave function. I. Gianturco,
Franco A., 1938-. II. Stefani, G. III. Series. QC793.5.E628W38 1984 539.7'2112 84-10578
ISBN 0-387-13347-X (U.S.)

Printing and binding: Beltz Offsetdruck, Hemsbach/Bergstr.
2152/3140-543210

FOREWORD

The present Volume of Lecture Notes in Chemistry fulfils one of
the stated aims of the Series, that of disseminating results discussed
and evaluated at recent scientific international conferences; in our
case a Satellite Meeting of the well-known Conference Series on the
Physics of Electronic and Atomic Collisions, the XIIIth ICPEAC, which
took place in Castelgandolfo, near Rome, from 23 to 25 July 1983.

Since the Satellite Meeting attracted a widely international and in-
terdisciplinary audience whose general consensus was one of warm appro-
val for the scientific level achieved during it, we hope that the present collection of essays will be met by similar success, thus warran-
ting our having asked the participants to work still further for us.

Before turning to their efforts, however, it is only just to thank
the Italian National Research Council (Chemistry Committee and Physics
Committee), the University of Rome, the C.N.R. Institute M.A.I. of the
Rome Research Area (Montelibretti) and the E.N.E.A. Organisation for
their financial aid, which made the Castelgandolfo Meeting possible.

We warmly acknowledge the professional expertise of the staff at
Villa Montecucco and for their collaboration we are grateful to: Rita
Abbasciano, Catherine Cajone, Lucilla Crescentini, Roberta Fantoni, An-
tonio Montani, Amedeo Palma, Rosario Platania, Maurizio Venanzi.

A special word of gratitude is owed to Dr. Rossana Camilloni for her
dedication and selfishness throughout the planning and execution of the
Meeting, Anna Giardini-Guidoni, as lively and active as usual as Conference Cochairman, is also warmly thanked.

F.A. Gianturco
G. Stefani

Rome, Fall 1983

ABSTRACT

In recent years the development and rapid advancements in the experimental detection of scattering events involving electron beams and atomic or molecular targets in the gaseous phase have produced a great wealth of highly accurate data of very detailed features with regard to the dynamical outcomes.

As a result of this experimental challenge, many attempts have been made to produce theoretical models of increasing reliability, and to unravel in ever greater detail the specific forces that preside over the essentially manybody nature of the relevant interactions.

The present volume, therefore, aims at bringing together under the same editorial roof the large variety of experts, experimentalists and theoreticians, who have examined the different facets of what is essentially the same problem, often developping specific languages of limited diffusion when describing these apparently diverse experimental facts.

Thus, the various results of low-energy collisions, where the detailed dynamical features are of prime importance in explaining things, are discussed next to those related to electron impact excitation and/or ionisation of atoms and molecules, where instead the traditional wisdom often assumes that all can be understood in the main via an increasingly more precise knowledge of the target electronic structural features. As it turns out, both distinctions are somewhat of a simplification, since both theoretical and experimental evidence begins to show that one needs to know, at comparable levels of accuracy, both the target wavefunctions and the essential ingredients of the dynamical interactions.

Contents: Low energy electron-molecule scattering: the experimental findings. Low-energy electron-molecule scattering: the theoretical tools.

Coupling nuclear motion with the impinging electron: theoretical models. Scattering of high energy electrons by atoms and molecules: experimental studies. Molecular excitation by electron impact: the theoretical treatment. The (e,2e) reactions on atoms and molecules: experiments and theoretical comparisons.

CONTENTS

PART I   *LOW ENERGY ELECTRON-MOLECULE SCATTERING:*

              THE EXPERIMENTAL FINDINGS.

PART II *LOW ENERGY ELECTRON-MOLECULE SCATTERING:*
THE THEORETICAL TOOLS.

PART V
MOLECULAR EXCITATION BY ELECTRON IMPACT:
THE THEORETICAL TREATMENTS.

PART VI
(e,2e) REACTIONS FOR ATOMS AND MOLECULES:
EXPERIMENTS AND THEORETICAL COMPARISONS.

INTRODUCTION

The contributions to this volume represent one of the several positive outcomes of an International Conference on the same subject which took place in Castelgandolfo, near Rome, in the summer of 1983.

The aim of the conference was to enable scientists, who usually practice their art in different and separate areas of research, to compare notes as to the similar and complementary aspects of their findings with regards to atomic and molecular structures. Initially the meeting appeared to all simply as topical and stimulating; but in fact it offered the opportunity of fastening several collaborations which were more specifically defined during the International Conference on the Physics of Electronic and Atomic Collisions that took place in Berlin immediately following the Castelgandolfo Meeting.

The diversity of the various experimental techniques described in the following Sections, and the differences between the specific jargons pertaining to each of the theoretical models which are also discussed here are only apparently so: instead they all cooperate in illustrating merely different facets of what is actually a complex interplay between structural features of atoms and molecules and the characteristics of the dynamical processes during which they are examined.

In the last 10-15 years, in fact, experimental progress has gathered a rapidly expanding body of detailed evidence on phenomena in which particle interactions play a more elaborate role than previously envisaged, and thus they exceed the scope of ordinary perturbation treatments where usually independent particle models are chosen and where the inclusion of particle interactions is done by lowest order perturbation theory.

Moreover it has often appeared reasonable to assume, in a qualitative way, that when considering inelastic collisional processes, the relative ratio between kinetic energy of the impinging electron and amount of energy transferred to atomic or molecular targets could conveniently be divided into two broad areas: (i) the slow-electron processes where the above ratio remains small and chere the key to the understanding of the physics involved lays in the unravelling of the many-body forces presiding over the dynamics rather than in detailed structural properties of the target atom or molecule, and (ii) the fast electron impact processes, where the ionic projectile acts mainly as a probe of the structural features of each target system and where the amount of energy transferred is only a small fraction of the total kinetic energy available to the electron. One might therefore expect to have, on one hand, collisional events which are dynamics-dominated and, on the other, those which are instead characterized in the main by the specific target structural properties. It is the aim of this volume of essays to show how both the sharpening of theoretical tools and the improved accuracy of the experiments have contributed in very recent years to blur the precise boundaries between these areas, hence indicating the fundamental similarity between quite different scattering events and the need for explaining both dynamics and structural properties at a comparable level of reliability.

When low-energy scattering events are considered, for instance, electron impact processes with atoms and molecules play a central role in a wide variety of naturally occurring and laboratory-produced phenomena. Usually mostly ground-state molecular targets and molecules in metastable excited states have been considered, due to the great importance in the study of ionospheric and auroral processes, of planetary atmospheres and of a variety of plasmas, just to cite a few examples. The corresponding cross sections, integral and differential, for individual rotovibrational transitions are increasingly found to depend not only on the range of energies examined but on the detailed way in which the electronuclear forces of a given target act on the light mass of the electron, i.e. how precisely the strong anisotropy of the bound electrons distribution guides the motion of the projectile in different spatial regions. As a consequence of recent theoretical results, such as those discussed in Parts II and III of this volume, it is becoming clearer that at low collision energies different spatial regions call for different representation of the dynamics and hence require the well-established use of different frames of reference in which to generate the continuum functions. At the same time, the accuracy of such functions and their ability to fit experimental data is found to be more sensitively dependent on the corresponding accuracy of the target wavefunctions employed than was previously thought to be the case, thus highlighting the need to devise accurate "structural" models to reach quantitative agreement with the recent wealth of ever more precise experiments such as those discussed in Part I.

The picture that seems to emerge from the present essays is one in which low energy collisional events constitute an essential tool for unravelling the mechanisms and getting a better insight on the largely unknown pathways through which many-electron systems (and many-nuclei systems) evolve from one stable (or metastable) state to another. Within this context, the various outcomes of the measurements can be qualitatively analysed in a remarkably unified fashion by articulating the corresponding theoretical techniques into subsections: here intermediate results can be interpreted step by step with the light mass of the electron playing an important role in deciding the dominant causes for each physical effect. Even when a more quantitative treatment needs to be used, the above partition turns out to be still very significant; however a more realistic, albeit more computationally demanding, description of the target structural features is definitely necessary, thus requiring an increasingly more effective interface with that large body of knowledge and expertise that deals with molecular structure calculations and which is generally known as computational quantum chemistry. This is even more evident when polyatomic systems are to be studied, as good experiments in this area increasingly indicate by making available the detailed resolution of specific inelastic channels for energy deposition and distribution among the molecular degrees of freedom.

At the other end of the scale, fast-electron impact spectroscopy has for a long time been considered a prime tool for elucidating the electronic structure of atoms and molecules, as witnessed by the numerous studies that have been carried out since the famous Franck and Hertz experiment of 1914, and which have aimed primarily at determining the energy level structure of the electronic bound states of the system being studied. Because of the energy of the impinging electrons, this structural information could be extracted within the framework of the first Born approximation thus generally bypassing the need to know in any greater detail the precise role played by the dynamics of the events. In an additional set of experiments now employing electron coincidence spectroscopy, one could also completely determine the kinematics in electron impact ionizations, or (e,2e) collisions, and use the measured (e,2e) differential cross sections to assess electron separation energies of the target as well as momentum profiles for the different transitions to the several final ion states. Obviously, for a clean knockout we must have high energy electrons, in order that the interaction of the incident and outgoing electrons with the potentials of the target and final ion are negligible, after which one can then treat the electron waves as plane waves. Moreover according to the choice of geometric arrangements which are used in doing the coincidence measurements, one can analyse events within the domain of a binary encounter situation (symmetric collisions, coplanar and noncoplanar) or within the range of asymmetric collisions. The latter collisions are of prime importance in studying the ionization reaction mechanism when low energies are used or, with highly energetic electrons, when sampling distant collisions that are sometimes referred to as dipole (e,2e) events in that they simulate photoelectron spectroscopy data for the system under study.

Today all the above experimental findings are achieving a high level of accuracy, as shown by the results discussed in Part IV and VI of this volume, an important consequence of which has been the increased awareness of the larger bearing that the dynamical mechanisms have on the types, features and number of observed events. It is interesting to note for example, that although scatteinrg resonances for atoms and molecules were already observed in the 30's, they were not well understood at the time. Now nearly all analogues to resonance types already known in nuclear physics have been detected in atomic and molecular physics, which means that electron impact experiments have become important tools for disentangling in a complementary way, the same specific forces which have already been detected within the low-energy range of collision energies. Moreover, the most detailed information available about the dynamics of single ionization reactions has been obtained by analysing triple differential cross sections, i.e. by measuring the probability that in a single ionization process an incident electron of energy E_o and momentum $\underline{K}_o$ will produce on collision with the molecule, two scattered electrons having prefixed values of energy and momentum and coming out at specific orientations with respect to the incident beam. The theoretical understanding of such observables is extremely challenging since it contains the full presence of many-body interactions coupled with the special features due to the infinite range of the Coulomb potential. Part V and VI of this volume thus show the essential unity that exists between the methods and tools employed to clarify high-energy impact processes and (e,2e) reactions, and those methods and techniques that deal with slow electron encounters with atoms and molecules previously analysed in Part I, II and III.

In conslusion, the unusual presentation of essays devoted to both low-energy and high-energy collision phenomena involving electrons and atoms or molecules, as offered in this volume, underscores our need to know at increasingly more comparable levels of confidence the dynamics of encounters and the detailed structural properties of the molecular targets. The hope is that this increased awareness will help practitioners in different subfields to transfer effectively, from one subfield to another, each specific body of reliable facts once they consider it common knowledge.

Rome, Fall 1983

F. A. Gianturco
G. Stefani

PART I

LOW ENERGY ELECTRON-MOLECULE
SCATTERING: THE EXPERIMENTAL
FINDINGS

EXPERIMENTAL RESULTS IN ELECTRON-MOLECULE

INELASTIC COLLISION EXPERIMENTS

Hiroshi Tanaka
Department of General Sciences
Sophia University
Chiyoda-ku, Tokyo 102, Japan

Introduction

In recent years, there has developed a great deal of interest in electron-polyatomic
molecule collision processes due to the following reasons: 1) considerable progress
in the development and computational treatments of electron-polyatomic molecule collisions, 2) continuing improvement of the energy resolution of electron energy-loss
spectroscopy, and 3) a demand for extensive quantitative data on electron molecule
collision cross sections for use in many related fields.

Electron-molecule collisions include a wide range of different processes: elastic
scattering, rotational, vibrational, and electronic excitations, electron capture (temporary negative ion state), ionization, dissociation, and their varios combinations.
At present, fragmentary cross section data are available for a number of linear molecules, but unfortunately, a complete set has been determined only for the nitrogen
molecule. The situation is much worse for electron-polyatomic molecule collisions.
However, their total electron scattering cross sections have been investigated extensively using electron-transmission and electron-swarm methods. But, up to now, a
breakdown into single processes has not been achieved in the electron beam experiments.
In this report, we summarize the normalized, absolute measurements of the differential
cross sections at low and intermediate impact energy for vibrationally elastic, rotationally inelastic, and vibrationally inelastic collisions for some of the molecules
of the paraffine series, especially methane. We also report the integral cross sections for individual excitation processes obtained by integrating the measured DCS's,
and compare the sum of each integral cross section with the total cross sections previously measured by different methods, as well as theoretical calculations.

Experimental Technique and Procedure

The apparatus[1] is of the conventional crossed molecular-beam and electron-beam type.
The overall resolution of the electrostatic cylindrical monochromator and analyzer
is 15 - 50 meV. Scattering angles are covered from 30° to 140°, and the energy range
can be adjusted from 3 to 30 eV.

The determination of the absolute cross section encounters difficult experimental
problems for which, at present, no completely satisfactory solution exists, and this

limits the accuracy of the cross section data: For elastic scattering DCS, i.e.,
vibrationally elastic scattering, the absolute values of the DCS were obtained from
measurements using the so called relative flow technique[2], which is based on a comparison with the known elastic DCS of helium[3]. For vibrational excitation, the ratios
of the peak heights of the $0 - \nu$ vibrational energy loss spectrum and the elastic
scattering spectrum was determined experimentally. To place the vibrational excitation DCS on the absolute scale, the ratios are then multiplied by the absolute elastic
DCS mentioned above. For rotational transitions, direct observation of the rotational
branches of methane would require a resolution better than 0.65 meV. However, broadening of the vibrationally elastic energy loss spectra can be observed even with a
resolution of 20 to 24 meV. The broadening is due to the superposition of unresolved
initial and final rotational states. We applied Read's high-J approximation[4] and by
a least square fitting, extracted the differential cross sections for pure rotational
transitions from the observed loss profiles.

The estimated errors are as follows: $\pm 15 - \pm 30\%$ for elastic DCS's, $\pm 35\%$ for elastic
integral cross sections, $\pm 25 - \pm 50\%$ for vibrational DCS's, and $\pm 40\%$ for vibrational
integral cross sections. They were obtained as the square root of the sum of the
squares of the statistical spread in the intensity, errors associated with the relative intensity measurements, normalization, and extrapolations.

Results and Discussion

a) Integral cross sections. Fig. 1 gives the complete set of cross sections for
methane which is one of the goals of our measurements. As can be seen, the major
contribution to the total cross section comes from elastic scattering. The cross
section for vibrational excitation is large, that is, vibrational excitation is quite
efficient(about 10^{-16} cm^2). But, the surprizing feature of this graph is the size of
the rotational excitation cross section. Although, in general, rotational excitation
cross section is weak, here it is comparable with the electronic and vibrational excitation cross sections. The electronic excitation cross section presented in this
figure is based on recent measurements of Vuskovic and Trajmar[5], and the ionization
cross sections are from Rapp et al[6]. In the same figure, we also compare the integral
cross section for elastic scattering calculated by Gianturco and Thompson[7] with our
results. The Ramsauer-Townsend minimum at low energies and the broad maximum at
higher energies near 8 eV are both correctly described. This calculation shows that
the single center expansion method with model exchange and polarization potential
yields relatively good results for non-linear molecules. It also shows that T_2, A_1,
and E are the most important scattering states, and predicted that the partial waves
for the T_2 wave function are $\ell = 1, 2, 3, 4,$ and 5, and that the shape resonance would
exhibit a strong $\ell = 2$ behaviour at 8 eV. In Fig. 1, two recent calculations of elastic scattering and rotational excitation cross sections by Abusalbi et al[8] and Jain
and Thompson[9] have been included.

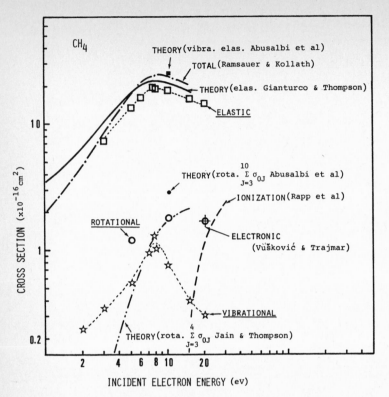

Fig. 1. Integral cross sections for vibrationally
elastic scattering(□), rotational excitation(○),
vibrational excitation(☆), electronic excitation
(+), ionization(--), and total scattering of electrons in CH_4. The various caluculated cross sections are shown by the indicated symbols.

b) DCS for vibrationally elastic scattering. In a previous report[10], we showed that
in elastic e + CH_4 scattering, the angular distribution near 8 eV is chracteristic of
a d-wave-dominated weak resonance as was theoretically predicted by Gianturco and
Thompson. In Fig. 2, our DCS at 10 eV are compared with the two recent calculations
mentioned above. The calculated DCS for the effective potential SEPlke of Abusalbi's
paper agrees well with our measurements over a wide range of scattering.

c) DCS for vibrational excitation. Another indication of a broad maximum near 8 eV
comes from the measurements of the vibrational excitation of the ground state of CH_4.
Methane belongs to the symmetry group T_d. There are in all five symmetry species in
three of which a normal vibration is possible, $A_1(\nu_1)$, $E(\nu_2)$, and $T_2(\nu_3, \nu_4)$, although
higher vibrational levels may also be found in A_2 and T_1. Due to the resolution of
the spectrometer, the ν_2, ν_4 modes and the ν_1, ν_3 modes overlap into composites ν_{24}
and ν_{13}. As shown in Fig. 1, near 8 eV, the cross sections of both ν_{24} and ν_{13} modes

are found to be considerably enhanced. In general, the coupling between electronic and vibrational motions induced by resonant scattering enhances the vibrational excitation relative to that induced by direct scattering. In order to gain further understanding of this resonance behaviour, we have investigated the angular distribution of the vibrational excitation. Fig. 3 shows that the angular distribution in both modes is characteristic of a d-wave near 7.5 eV. Methane has a ground state configuration $(1a_1)^2(2a_1)^2(1t_2)^6(3a_1)^0(2t_2)^0$ .... , yielding an a$^1A_1$ ground state. The lowest unoccupied orbital is $3a_1$, and the next unoccupied orbital is expected to be $2t_2$. In a recent report[11], we interpreted the broad resonance most probably as the $^2T_2$ compound state comprising the target molecule plus an electron in the $2t_2$ orbital.

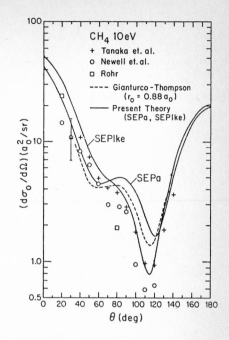

Fig. 2. Electronically and vibrationally elastic, rotationally summed DCS's as functions of the scattering angle for CH_4 (from Ref. 8).

We have also calculated the angular distributions with the help of the theory of angular correlation[12] with irreducible tensors for the T_2 molecular symmetry group of the compound state. Assuming the resonance to be of symmetry of T_2, as mentioned above, the resonance decays into the following modes of vibrations:

$$e + CH_4 \longrightarrow CH_4^{-*} \longrightarrow e + CH_4^*$$

$$A_1 \longrightarrow T_2 \longrightarrow A_1 \quad (\nu_1)$$

$$A_1 \longrightarrow T_2 \longrightarrow E \quad (\nu_2)$$

$$A_1 \longrightarrow T_2 \longrightarrow T_2 \quad (\nu_3, \nu_4).$$

For our resolution, the decaying processes corresponding to ν_{24} and ν_{13} have to be combined, namely A_1 and T_2 as well as E and T_2. In each process, the electron waves in the incident and exit channels are given by the direct product such as $A_1 \times T_2 = T_2$ in $A_1 - T_2$. Froma consideration of the large centrifugal barrier, higher partial waves have been excluded. Thus, we have fitted the observed angular distribution by assuming that a mixture of s, p, and d waves is sufficient to discribe the decay and obtained the curves as shown in Fig. 3. These fits are quite reasonable. Also, the partial contributions of the vibronic symmetries A_1, E, and T_2 to each composite mode are presented. The agreement between the measurements and the theoretical prediction is a further indication of a broad resonance near 8 eV. The shape resonance shows clearly an $\ell = 2$ behaviour.

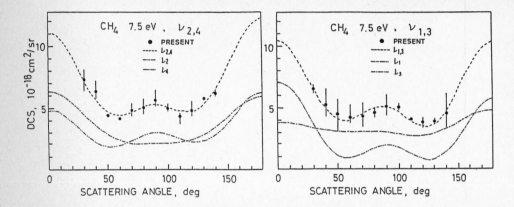

Fig. 3. Angular distributions for vibrational excitation of the ν_{24} and ν_{13} modes in CH_4 by electron impact at 7.5 eV. The points represent the measured angular distribution. Calculated angular distributions are shown by the dashed lines, and the partial contributions by the indicated lines.

d) DCS for rotational excitation[13]. Recent development in computational techniques has made possible fairly accurate semiempirical and ab initio calculations on rotational transitions of electron-methane collision[8,9]. Due to the required high energy resolution of the spectrometer, so far, no experimental information is available. However, there exists the possibility to extract state-to-state cross sections from a rotationally unresolved spectrum as proposed by Shimamura[14]. In part, this possibility did motivate us to perform these measurements.

Methane is a spherical top molecule and it has neither dipole nor quadrupole moment, that is, no dipole-type rotational transitions are to be expected, but higher multipole interactions have to be considered. For tetrahedral molecules, such as CH_4, the selection rule is $\Delta J = 0$, ± 1, ± 2, $\pm 3, \ldots$. The transitions $\Delta J = 0 - 1, 2$ are forbidden. Actually, due to centrifugal distortion effects, methane in its ground state has a tiny dipole moment (about 5.4×10^{-6} D). As mentioned above, we have measured the broadening of the spectra and obtained, from these measurements, the rotational branches with $\Delta J = 0$, ± 1, ± 2, ± 3, and ± 4 by the high-J approximation: The intensity profile for rotational transition from J_i to J_f at scattering angle θ is given by

$$I(E_0, \theta, T) \propto \int dE\, f(E, E_0, \theta) \sum_{\Delta J} P(E, \Delta J, T)$$

$$\times \sigma(|\Delta J|, \theta)[(2J_f + 1)/(2J_i + 1)], \qquad (1)$$

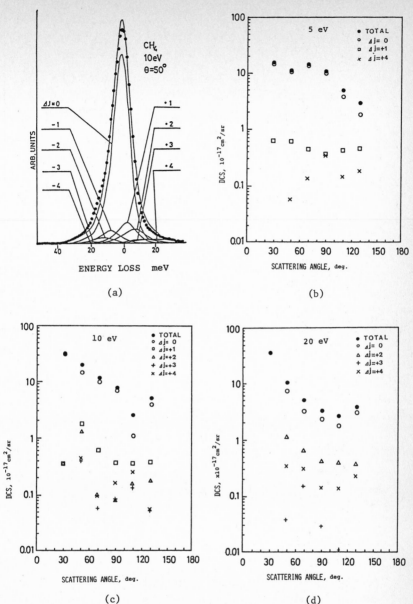

Fig. 4. Line-shape-analysis for the energy loss profile at 10 eV for scattering angle of 50°(a). Differential cross sections of pure rotational excitations: (b) 5 eV, (c) 10 eV, and (d) 20 eV.

where σ is the cross section depending only on ΔJ and θ, f is the apparatus profile determined from helium, and P is the population density distribution. The branching ratios of the pure rotational transitions were obtained from a line-shape analysis, i.e., a least square fitting of this equation with 5 parameters ($|\Delta J| = 0, 1, 2, 3, 4$) to the experimental data as shown in Fig. 4. Here, the sum of the rotational cross sections was normalized to the rotationally unresolved cross section, that is, to the vibrationally elastic DCS's mentioned above (b). In the energy range of 5 to 20 eV and for scattering angles of 30° to 130°, the main components are the branches $\Delta J = \pm 1$, ± 4 at 5 eV, $\Delta J = \pm 1, \pm 2, \pm 3, \pm 4$ at 10 eV, and $\Delta J = \pm 2, \pm 3, \pm 4$ at 20 eV. Transition $\Delta J = \pm 4$ appears at all energies investigated while transitions $\Delta J = \pm 1$ have been observed only at 5 and 10 eV, but not at 20 eV. Transitions of $\Delta J = \pm 2$ and ± 3 do not appear in the spectrum at 5 eV. These are only preliminary results and need further confirmation. For the same reason, an error estimation has been ommitted. Also note that, in the high-J approximation, the differential cross section for a fixed ΔJ is assumed to be independent from J_i and forms a kind of average over the whole rotational distribution. As discussed in the recent papers of Abusalbi et al and Jain and Thompson, the short range interaction is more important than the point-octupole interaction so that the effect of polarization can not be neglected. Their integral cross sections have been included in Fig. 1. But, here, we can not make a direct comparison with our data, because their calculations refer to a state-to-state rotationally inelastic cross section. To extract state-to-state cross sections by using Shimamura's method, the measurements must be refined and probably repeated many times to decrease the errors. A good comparison of experimental and theoretical results would be extremely desirable.

Conclusion

To summarize. 1) A reasonably complete set of experimental cross sections for methane was determined for the first time with an electron-beam experiment, 2) a broad maximum can be found near 8 eV and represents a weak shape resonance, 3) it seems that the pure rotational excitation is large and comparable with the vibrational and electronic excitation cross sections.

This work is based on a collaboration with Dr L Boesten, Dr I Shimamura, and Mr N Onodera.

References

1. H. Tanaka, T. Yamamoto and T. Okada, J. Phys. B: At. Mol. Phys. 14 2081 (1981)
2. S. K. Srivastava, A. Chutjian and S. Trajmar, J. Chem. Phys. 63 2659 (1975)
3. D. F. Register, S. Trajmar and S. K. Srivastava, Phys. Rev. A21 1134 (1980)
4. F. H. Read, J. Phys. B: At. Mol. Phys. 5 255 (1972)
5. L. Vuskovic and S. Trajmar, J. Chem. Phys. 78 4947 (1983)
6. D. Rapp and P. Englander-Golden, J. Chem. Phys. 43 1464 (1965)
7. F. A. Gianturco and D. G. Thompson, J. Phys. B: At. Mol. Phys. 9 1383 (1976)
   F. A. Gianturco and D. G. Thompson, J. Phys. B: At. Mol. Phys. 13 613 (1980)
   S. Salvini and D. G. Thompson, J. Phys. B: At. Mol. Phys. 14 3797 (1981)
   A. Jain and D. G. Thompson, J. Phys. B: At. Mol. Phys. 15 L631 (1982)
8. N. Abusalbi, R. E. Eades, T. Nam, D. Thirumalai, D. A. Dixon and D. G. Truhlar, J. Chem. Phys. 78 1213 (1983)

9. A. Jain and D. G. Thompson, J. Phys. B: At. Mol. Phys. <u>16</u> 3077 (1983)
10. H. Tanaka, T. Okada, L. Boesten, T. Suzuki,T. Yamamoto and M. Kubo, J. Phys. B: At. Mol. Phys. <u>15</u> 3305 (1982)
11. H. Tanaka, M. Kubo, N. Onodera and A. Suzuki, J. Phys B: At. Mol. Phys. <u>16</u> 2861 (1983)
12. F. H. Read, J. Phys. B: At. Mol. Phys. <u>1</u> 893 (1968)
    D. Andrick and F. H. Read, J. Phys. B: At. Mol. Phys. <u>4</u> 389 (1971)
13. H. Tanaka, N. Onodera and L. Boesten, 13th ICPEAC (1983, Berlin), Abstract p245
14. I. Shimamura, Chem. Phys. Lett. <u>73</u> 328 (1980)

LOW ENERGY ELECTRON SCATTERING

IN ORGANIC MOLECULES

Michael ALLAN
Institut de chimie physique
de l'Université

CH-1700 Fribourg, Switzerland

Our present knowledge of low energy electron molecule scattering is concentrated on small, often diatomic or triatomic molecules, because of their simplicity, low number of vibrations and the possibility to understand the scattering in terms of a more rigorous theory. This study aims at electron scattering with larger polyatomic molecules. The motivation for this extension is twofold. 1. Qualitatively new scattering phenomena might be found, not possible in small molecules. The large molecule represents a link between the small molecule and bulk matter. 2. The information obtained, in particular the energies and dynamics of the anion states and the energies of triplet states and other states not accessible by light absorption, has important applications in understanding the chemistry of these molecules.

The present contribution presents two aspects of electron-polyatomics collisions: Properties of the long lived anion formed at non-thermal energies in p-benzoquinone are studied by time resolved energy loss spectroscopy and unspecific excitation of a vibrational quasi-continuum near threshold is found in several polyatomic molecules.

Experimental

The transmission and energy loss spectra were recorded using a trochoidal electron inpact spectrometer described recently [1]. An electron beam with a narrow energy spread collides with a static gaseous sample. Electrons inellastically scattered at $0°$ and $180°$ are energy analysed by two trochoidal monochromators in series and detected. Spectra, where the analyser energy was not constant, are corrected for the analyser transmission function. In time resolved experiments the electron beam was pulsed and the delayed coincidence technique was used to detect the electrons.

Long lived anion of p-benzoquinone

This long lived anion was first reported by Christophorou (1969) [2]. It is exceptional in two respects. First, metastable parent ions are usually formed only by impact of thermal electrons (e.g. SF_6^-). Second, unless coupling to a large density of states is assumed, its long lifetime would implicate an extremely narrow width which would preclude its observation in an electron impact experiment. In the present time resolved energy loss experiment [3] the negative ion autodetaching with a lifetime in the μs range was observed in collisions of 1.4 - 2.3 eV electrons with p-benzoquinone. Two decay

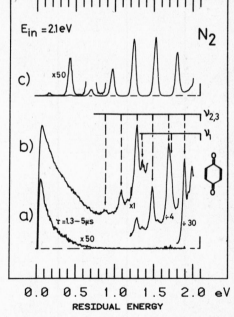

Fig. 1 Spectra of electrons detached following a capture of a 2.1 eV electrons. In the curve a) only electrons autodetached with a time delay within the specified interval were registered.

times are found following the capture of a 2.1 eV electrons. First a rapid electron detachment with $\tau < 0.07$ μs (derived from direct measurement) and probably with $\tau < 10^{-14}$ s (deduced from the absence of vibrational structure), followed by a slow autodetachment with $\tau > 1$ μs. Time resolved energy spectra of the detached electrons (fig. 1) reveal that the fast autodetachment leads primarily to excitation of discrete vebrational levels with low vibrational quantum number (curve b). In contrast, after the slow autodetachment nearly all of the incident energy is transfered to nuclear excitation of the target molecule and only near zero energy electrons are observed (curve a). These observations can be understood in terms of two rapid competing processes following the capture of a 2 eV electron. They are a rapid autodetachment on the $10^{-14} - 10^{-15}$ s timescale and a rapid radiationless transition leading to a vibrationally excited species which then undergoes a slow vibrational autodetachment. The radiationless transition could involve a nuclear rearrangement to a new metastable structure of the anion, somewhat analogous to the metastable bent CO_2^-.

Near threshold vibrational excitation

A noteworthy feature in the curve b in fig. 1 is the excitation of a vibrational quasi-continuum with peak at threshold. This observation is independent of the long life-time ion. A survey of inelastic processes recorded at fixed energies above threshold is shown in fig. 2. The spectrum with 20 eV residual energy is dominated by electronic excitation. In contrast, the strongest peaks in the threshold spectrum are below the

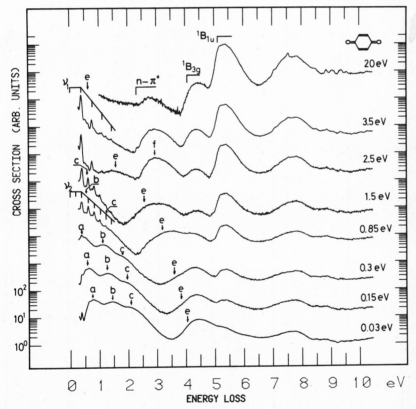

Fig. 2. A series of electron energy loss spectra in p-benzoquinone, recorded with constant residual energies indicated on the right. The spectra are shown on a semilog scale to accomodate the large variations in signal intensity. The upper spectra are offset by arbitrary amounts. The lower case characters with arrows indicate points with incident energies corresponding to resonances.

energy of the lowest electronic state and must correspond to pure vibrational excitation. At intermediate residual energies both electronic and vibrational excitation are important. Although near threshold very intense vibrational losses up to 3 eV are observed, no structure due to individual vibrational levels is visible, except some weak structure at energy losses below 0.8 eV. Instead, broad peaks appear which correspond to the shape resonances, whose energies were determined in the transmission spectrum [3]. With increasing residual energy the points of fixed incident energy move to the left on the energy loss scale, and the broad peaks also move to the left, in a manner consistent with their assignment to resonances, which are fixed on the incident energy scale. Whenever an incident energy corresponding to a shape resonance reaches the left end of the spectrum, a long series of vibrational peaks appear. These observations can be summarized in terms of two different types of vibrational excitation.

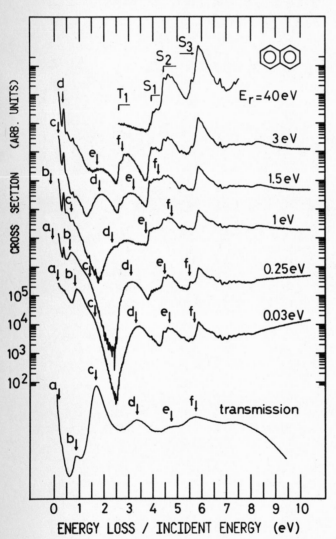

Fig. 3. Electron energy loss spectra in naphthalene, presented as in fig. 2. The bottom curve, a transmission spectrum, shows the approximate total scattering cross section, on a linear scale, with zero offset and slightly tilted to improve the visibility of the resonances.

First, a specific excitation of low quanta of only a few vibrations is observed at the energies of shape resonances. Some resonances excite mainly the C=O stretch, others the C-H stretch vibrations, and this selectivity is related to the electronic structure of the resonances. A second pattern in vibrational excitation is an unspecific excitation of a vibrational quasicontinuum where no structure due to individual vibrations can be discerned. This second kind of vibrational excitation peaks at threshold, high vibrational energies are excited preferably, and the excitation is also enhanced at energies of the shape resonances. A detailed account of this work will be published elsewhere [4].

A similar pattern in vibrational excitation was also found in naphthalene, illustrated in fig. 3. Six resonances are discernible in the transmission spectrum, in agreement with previous results [5]. Five of these, a - e, are seen to cause excitation of vibrational quasicontinuum at various energies above threshold. Fig. 4 shows energy dependence of the vibrational excitation for three different energy losses. The role of the resonances b - e is apparent. A threshold peak is observed in the botton curve, where the resonance b is active at threshold.

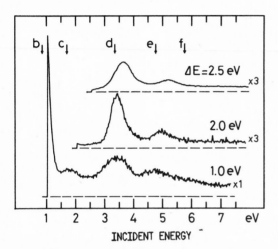

Fig. 4. Energy dependence of the indicated vibrational energy losses in naphthalene.

The last molecule discussed will be diacetylene (1,3-butadiyne). The transmission spectrum in fig. 5 reveals two peaks. The lower at 1.0 eV has a weak structure (.093 eV) and corresponds to the lowest π^* $^2\Pi_u$ resonance. The higher around 5.6 eV probably contains overlapping contributions of the second π^* $^2\Pi_g$ resonance as well as a number of σ^* and doubly excited states. The 1 eV resonance causes intense vibrational excitation of the $n\nu_2$ (C≡C stretch) and the $2\nu_{6,8} + n\nu_2$ progressions (ν_6 and ν_8 are nearly degenerate C-H bend vibrations). Near threshold a large number of peaks corresponding to additional excitation of low frequency vibration appear and congest the spectrum. The diacetylene molecule is smaller than the two molecules discussed above and might represent an intermediate case between small and large molecule behaviour. Many more vibrations are excited at threshold than further off threshold, but because of the smaller number of vibrations present in the molecule the individual vibrations are still resolved at $E_r = 0.1$ eV. The intensity of the excitation of the low frequency vibrations is less than in the large molecules and no threshold peak is observed for them.

Unspecific vibrational excitation near threshold was also observed in benzene [1]. This phenomenon appears to be present in many, perhaps most large molecules. It is probably responsible for the appearance of shape resonances in many earlier trapped electron spectra of large molecules.

This work is part of project No 2.422-0.82 of the Swiss National Science Foundation.

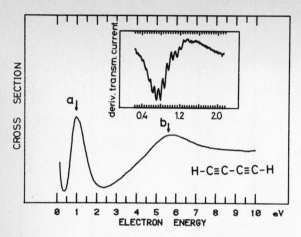

Fig. 5. Electron transmission spectrum of 1,3-butadiyne.

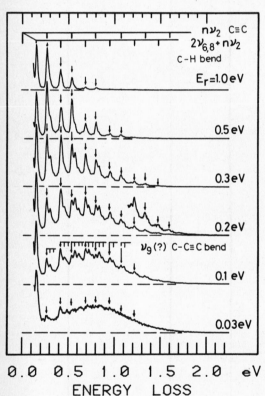

Fig. 6. Vibrational excitation in 1,3-butadiyne.

References

[1] M. Allan, Helv. Chim. Acta 65, 2008 (1982).

[2] L.G. Christophorou, J.G. Carter, and A.A. Christodoulides, Chem. Phys. Lett. 3, 237 (1969).

[3] M. Allan, Chem. Phys. (1983), in print.

[4] M. Allan, Chem. Phys., submitted for publication.

[5] K.D. Jordan and P.D. Burrow, Acc. Chem. Res. 11, 341 (1978) and private communication.

ELECTRON AFFINITIES OF ORGANIC AND ORGANOMETALLIC COMPOUNDS DETERMINED BY MEANS OF ELECTRON TRANSMISSION SPECTROSCOPY.

Alberto Modelli

Istituto Chimico "G. Ciamician", Università di Bologna,via F. Selmi 2, 40126 BOLOGNA, Italy

Giuseppe Distefano, Maurizio Guerra and Derek Jones

Istituto dei Composti del Carbonio contenenti Eteroatomi, C.N.R., via Tolara di sotto, 89, 40064 OZZANO E. (BO), Italy.

An electron with appropriate energy and angular momentum can be temporarely trapped into a normally unoccupied molecular orbital (MO). At large separation,the electron-molecule interaction is mainly due to an attractive polarization potential and to a repulsive centrifugal potential (associated with the angular momentum of the electron), whose combination gives generally rise to a barrier[1]. The typical life-time of these anions is of the order of 10^{-12}-10^{-15}s and their formation produces sharp variations (referred to as "resonances") in the electron scattering cross section. One of the most powerful techniques for the detection of resonances in gas-phase samples is Electron Transmission Spectroscopy (ETS)[2], in the format devised by Sanche and Schulz[3].

In the normal mode of operation, only the unscattered part of the electron current is transmitted to the collector and a signal is derived which is associated with the total scattering cross section. Besides, with the same apparatus, also a signal associated with the "back-scattering"cross section can be obtained[4]. In order to enhance the variations in the cross section due to resonant scattering, the derivative of the transmitted current is recorded as a function of electron energy.

The energy at which a resonance occurs, or attachment energy (AE), depends on the energy of the empty orbital into which the electron is temporarily captured and closely approximates the negative of the corresponding electron affinity (EA). ETS is therefore complementary to the more well-known photoelectron spectroscopy, which provides ionization energies. A limitation of ETS is that positive EA's, that is, those associated with stable anion states, cannot be measured. Moreover, electron addition to orbitals with pure "s" character ($\ell=0$) produces anions with too short a life-time to be detected. However, resonances associated with σ^* MO's can be observed, as in the case of the chloroethylenes[5] and of the chloromethanes[6].

ETS studies of the anion states of organic molecular systems are relatively recent (see reference 1 for a review) and only a very few metal complexes[7,8,9,10] have been analyzed up to date.

We employed ETS to investigate the stabilizing effect on Π anion states exerted by substituents containing a second row element (Si[11,12]
or S[13]) in comparison with the effect produced by substituents containing the corresponding first row element (C or O).

Here, we present some additional results on planar and rotated oxygen and sulphur benzene derivatives and on the transition metal complexes dicyclopentadienyl-iron, -chromium and -vanadium.

Fig.1 shows that in the planar phenol and anisole the interaction between the oxygen lone pair (l.p.) and the ring removes the degeneracy
of the empty benzene $e_{2u}(\pi^{*})$ MO, the $^{2}B_{1}$ anion state being destabilized
by about 0.5 eV. The anion state associated with electron addition to
the $5b_{1}(\pi^{*})$ MO, higher in energy and with a smaller wave function coefficient at the carbon atom attached to the substituent, is less destabilized.

In the *tert*-butoxy benzene derivative, where steric hindrance causes
a rotation around the $O-C_{ring}$ bond, the l.p./π^{*} interaction is largely
reduced. Consistently, the ET bands associated with electron capture into the $2a_{2}$ and $4b_{1}$ (π^{*}) MO's are not resolved and the value of 1.4 eV
is to be considered an upper limit to the AE into the latter MO. In add.
tion, the anion state associated with the $5b_{1}$ MO is stabilized by about

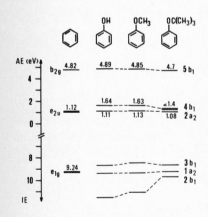
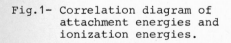

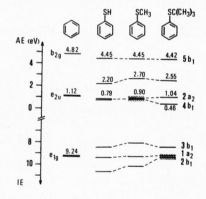

Fig.1- Correlation diagram of
attachment energies and
ionization energies.

Fig.2- Correlation diagram of
attachment energies and
ionization energies.

0.15 eV.

At variance with the oxygen analogues, in the planar -SH and $-SCH_3$ benzene derivatives, the $^2A_2$ and $^2B_1$ anion states are very close in e-nergy (see Fig.2) and give rise to only one unresolved band (at 0.2-0.3 eV lower energy than in benzene). Moreover, the higher-lying Π anion sta-te is stabilized by about 0.4 eV with respect to the $^2B_{2g}$ anion state of benzene.

These findings indicate that in the sulphur derivatives an additio-nal interaction between the ring π^* MO's and higher-lying empty orbitals of the substituent is present, which overcomes the effect of the l.p./ /π^* interaction. Mixing of the empty 3d orbitals of sulphur with the π^* MO's can account for the observed stabilization of the Π anion states in thiophenol and thioanisole. If the planar conformation of the neutral molecules were distorted upon anion formation, however, also σ^*_{C-S}/π^* mi-xing would be symmetry allowed and could play some role. To get better insight, we measured the AE's of the rotated $-SC(CH_3)_3$ derivative (see Fig.2), where σ^*_{C-S}/π^* interaction is certainly possible. As expected be-cause of the reduced l.p./π^* interaction, the anion state associated with electron capture into the $4b_1$ MO is stabilized. However, some con-tribution to this stabilization, together with a concomitant destabili-zation of the highest-lying $^2B_1$ anion state, can be ascribed to intera-ction with the anion state at 2.55 eV, probably associated with a σ^*_{C-S} orbital.

The difference between the ET spectrum of the rotated $-SC(CH_3)_3$ deri-vative and those of the -SH and $-SCH_3$ derivatives indicates that the a-nion states of the latter are planar. The ETS data are therefore consi-stent with an important role played by $3d/\pi^*$ mixing in the stabilizati-on of the Π anion states of the sulphur derivatives.

Fig.3 is a qualitative scheme of the frontier MO's of the dicyclope-ntadienyl transition-metal complexes (Cp_2M).

In the closed-shell Cp_2Fe, the lowest unoccupied MO is the $e_{1g}(d)$ orbital. The ET spectrum of Cp_2Fe (see Fig.4) displays two bands, ce-ntred at 0.63 and 2.74 eV. In agreement with the results of MS Xα cal-culations, the former is ascribed to electron addition to the $e_{1g}(d)$ or-bital and the latter to electron addition to both the e_{2u} and $e_{2g}(\pi^*)$ ring orbitals (calculated energy separation = 0.16 eV).

In the open-shell Cp_2Cr (e_{2g}^3, a_{1g}^1) and Cp_2V (e_{2g}^2, a_{1g}^1), addition of one electron to each empty orbital is expected to produce more than one anion state. Spin-spin coupling will be larger for electron capture in-to MO's localized at the metal center (where the unpaired electrons are localized). Electron capture into the $e_{1g}(d)$ orbital of Cp_2Cr gives ri-

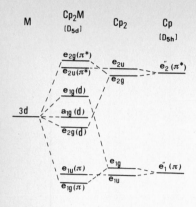

M   Cp$_2$M $[D_{5d}]$ Cp$_2$ Cp $[D_{5h}]$

$e_{2g}(\pi^*)$
$e_{2u}(\pi^*)$ — e_{2u}
$e_2''(\pi^*)$
e_{2g}
$e_{1g}(d)$
3d — $a_{1g}(d)$
$e_{2g}(d)$
$e_{1u}(\pi)$ — e_{1g}
$e_1''(\pi)$
$e_{1g}(\pi)$ — e_{1u}

Fig.3- Qualitative scheme of the frontier MO's of dicyclopentadienyl transition-metal complexes.

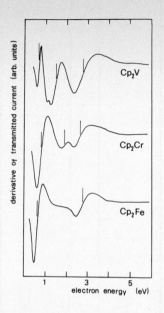

Fig.4- ET spectra of ferrocene, chromocene and vanadocene.

se to two bands (0.80 and 1.90 eV) and the four anion states (calculated energy spread=0.25 eV) associated with electron capture into the two π^* MO's give rise to the broader band centered at 2.63 eV. The Xα calculations predict the anion states associated with electron capture into the half filled a_{1g} and $e_{2g}(d)$ orbitals to be stable, and therefore not accessible in ETS.

According to the calculations, a different assignment is proposed for the three resonances of the spectrum of Cp$_2$V (0.68, 1.48 and 2.69 eV). The first resonance is ascribed to an anion state with half-filled $a_{1g}(d)$ and $a_{1g}(4s)$ character, the second resonance to the $^5E_{1g}$ anion state arising from electron capture into the $e_{1g}(d)$ empty orbital and, finally, the third broad resonance to the corresponding $^3E_{1g}$ triplet state and to the four Π anion states.

REFERENCES

1) K.D. Jordan and P.D. Burrow, Acc. Chem. Res., $\underline{11}$ (1978) 341.

2) G.J. Schulz, Rev. Mod. Phys., $\underline{45}$ (1979) 378, 423.

3) L. Sanche and G.J. Schulz, Phys. Rev., $\underline{A5}$ (1972) 1672.

4) A.R. Johnston and P.D. Burrow, J. Electron Spectrosc. elat. Phenom., $\underline{25}$ (1982) 119.

5) P.D. Burrow, A. Modelli, N.S. Chiu and K.D. Jordan, Chem. Phys. Letters, $\underline{82}$ (1981) 270.

6) P.D. Burrow, A. Modelli, N.S. Chiu and K.D. Jordan, J. Chem. Phys., $\underline{77}$ (1982) 2699.

7) J.C. Giordan, J.H. Moore and J.A. Tossel, J. Am. Chem. Soc., $\underline{103}$ (1981) 6632.

8) J.C. Giordan, J.H. Moore, J.A. Tossel and J. Weber, J. Am. Chem. Soc., $\underline{105}$ (1983) 3431.

9) A. Modelli, A. Foffani, M. Guerra, D. Jones and G. Distefano, Chem. Phys. Letters, $\underline{99}$ (1983) 58.

10) P.D. Burrow, A. Modelli and K.D. Jordan, in preparation.

11) A. Modelli, D. Jones and G. Distefano, Chem. Phys. Letters, $\underline{86}$ (1982) 434.

12) A. Modelli, G. Distefano, D. Jones and G. Seconi, J. Electron Spectrosc. Relat. Phenom., $\underline{31}$ (1983) 63.

13) A. Modelli, G. Distefano and D. Jones, Chem. Phys., $\underline{73}$ (1982) 395.

SHAPE RESONANCES IN ELECTRON - POLYATOMIC MOLECULES COLLISIONS

Michel TRONC, Laurence MALEGAT

Laboratoire de Chimie Physique
Université Pierre et Marie Curie
11, Rue Pierre et Marie Curie   75231   PARIS Cedex 05   France.

Shape resonances have been shown to play a prominent role both in partial photoionization cross sections and in vibrational inelastic differential cross sections.[1,2] Moreover a clear connection has been established between the shape resonances observed in K-shell photoabsorption (photoionization) and in vibrational inelastic D.C.S. of the first row diatomic molecules : the same manifold of shape resonances with the same symmetries, the same relative energies, and the same partial waves is observed for the two processes namely a π resonance at low energy (0-5 eV in electron scattering) and a σ resonance at higher energy (10-25 eV)[3]

The incoming electron, in the electron scattering process, $(e+M_{v=0} \rightarrow M^- \rightarrow M_{v \neq 0}+e)$ and the departing electron, in the photoionization process $(h\nu +M \rightarrow M^++e)$ can be trapped by their own centrigugal barrier at the periphery of the molecule : the strength of the attractive portion of the interaction potential determines the energy of the molecular compound state (M^- or M) lying in the electron-molecule (or electron - positive ion) continuum, while the lifetime of the resonance depends on the height of the barrier (related to the electron angular momentum) and on its width at the resonance energy.

Recent discrete basis set calculations[4] and multiple scattering continuum wavefunctions [5] have established a direct connection between the unoccupied antibonding (virtual) orbitals of the neutral molecule and continuum wavefunctions of unbound electrons at resonance energies. So that a simple picture emerges for the description of shape resonances in electron scattering and photoionization processes, in which each unoccupied valence orbital, with its specific charge distribution and hence its given symmetry, determines the angular momentum components of the impinging (escaping) electron which are active in its own trapping. A similar description would be to say that only the partial waves matching the spatial charge distribution of the unoccupied orbital tunnel through the barrier at the resonance energy.

In the N_2 molecule, for example, the cross section for excitation of the vibrational level v =1 shows two shape resonances (fig.1) at 2.3 and 22.5 eV

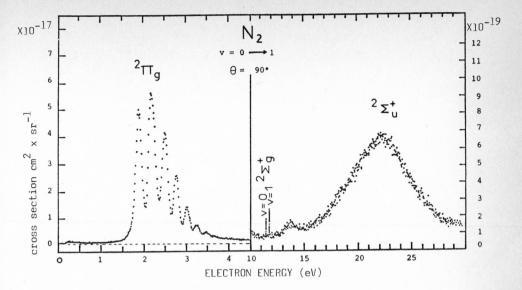

Figure 1 - Differential cross section for excitation of the v=1 level of the $X^1\Sigma_g^+$ ground state of N_2 at 90° in the 0- 30 eV energy range, showing the two shape resonances at 2·3 and 22.5 eV with trapping of the electron in the $1\pi_g$ and $3\sigma_u$ unoccupied valence orbitals.

- Figure 2 -

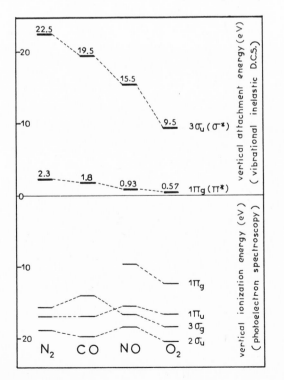

Correlation diagram of the valence occupied and valence unoccupied orbitals for the first row diatomic molecules.

The vertical ionization energies are obtained from photoelectron spectroscopy (TURNER et al Ref 8) and the vertical attachment energies from the differential cross sections for resonant vibrational excitation (TRONC et al Ref 3) with trapping of the extra electron in the antibonding $1\pi_g^*$ and $3\sigma_u^*$ orbitals respectively. The active partial waves extracted from angular distribution measurements are the d wave (l=2, m=±1) at 2.3 eV, the f wave (l=3, m=o), plus a contribution of the p wave (l=1, m=o) at 22.5 eV.

Similar observations in the first row diatomic molecules CO, NO, O_2 have been used to construct the correlation diagram of the unoccupied valence orbitals shown in figure 2, where unoccupied orbitals are marked by their vertical electron attachment energy. Such a correlation diagram is the counterpart for the valence unoccupied MO's, of the correlation diagram for the occupied MO's obtained from the vertical ionization energies [8].

The observation of shape resonances in polyatomic molecules and the extension of the simple picture mentioned above are of interest because these molecules have more unoccupied valence orbitals and more modes of vibration with different symmetries so that selective vibrational excitation can be observed [16]. Even if for complicated molecules with low symmetry, or for high energy orbitals the simple description is less valid, because of the greater number of angular momenta needed to describe the scattering (or to expand the unoccupied orbital charge distribution in the spherical harmonics) one can expect a qualitative picture of the process.

The vibrational inelastic differential cross sections (D.C.S.) appear as the most sensitive way to observe all the shape resonances because: i) their observation is not restricted by the dipole selection rules as it is the case in photoionization, ii) they can produce enhancement of the inelastic D.C.S.) clearly observable on a small monotonically varying background of direct excitation which is not always the case in photoionisation where autoionisation, interference effects and the presence of strong non resonant channels make their detection more difficult.

The electron transmission spectroscopy technique [6] (E.T.S.) has given characteristics of temporary anion states associated with low-lying π^* orbitals in many unsaturated molecules, but was innefficient to detect very short lived resonant states such as the one resulting from the trapping of electron in σ^* antibonding orbitals at intermediate energy. Moreover the E.T.S. gives no indication on the partial waves.

To generalize the conclusions from the observations made in the diatomic molecules, the acethylene molecule C_2H_2 was a good candidate because it is linear in its ground state, and isoelectronic with N_2, but at the same time it has different nuclei, a different geometry and a different molecular field. The LCAO MO calculations of the C_2H_2 ground state [7] show a $1\pi_g$ and a $4\sigma_u$ unoccupied valence orbitals (related to the $1\pi_g$ and $3\sigma_u$ of N_2) which are antibonding between the two carbon atoms, and a $4\sigma_g$ and a $3\sigma_u$ orbitals mainly antibonding between the carbon and hydrogen atoms (without equivalence in N_2) (figure 3). All these unoccupied orbitals have been detected as discrete states or continuum states (resonances) in photoabsorption and photoionization spectra, and they can be expected to produce shape resonances in electron scattering processes for energies of impinging electron roughly 10 eV higher than the kinetic energy of the ejected electron in the photoionization process, owing to the less attractive coulomb potential in M+e compared to M^++e. (table 1)

Figure 4 shows energy loss spectra of C_2H_2 at energies of the three broad shape resonances we have observed : at 2.6 eV the excitation is dominated by the ν_4 (ν_5) unresolved bending mode, the ν_2 stretching carbon-carbon

unoccupied valence orbital	ENERGY (eV)	SHAPE
4 σ_u	32.53	
4 σ_g	13.18	
3 σ_u	9.60	
1 π_g	6.83	

- Figure 3 -   Characteristics of the unoccupied valence orbitals of $C_2 H_2$ from the L.C.A.O.   M.O. calculations (Réf 7).

MO's	character	Vertical I.P. (eV)	Unoccupied MO's			
			1 π_g	4 σ_g	3 σ_u	4 σ_u
1 σ_g	C 1s n.b.	291.1				307[a]
1 σ_u	C 1s n.b.	291.1	285.9	288.1[a] +threshold		
2 σ_g	C-C b. / C-H n.b.	23.9			20.85[e] +continuum	
2 σ_u	C-C a.b. / C-H b.	18.8	15.31[b]	17.68[e] +continuum		
3 σ_g	C-C b. / C-H b.	16.4			13.33[b] +threshold[d]	
1 π_u	C-C b.	11.4	9.27[c]	9→10[c]		

- Table 1 -   Transitions from occupied to unoccupied valence orbitals in $C_2 H_2$.
a) K.Shell energy loss (Ref 9)
d) Photoionization mass spectrometry (Ref 10)
c) Excitation (Ref 12)
d) Photoionization (Ref 11)
e) Hartree - Fock photionization calculation (Ref 13).

mode, and a combination of $\nu_2 + \nu_4$; at 6.2 eV the excitation of ν_1 (and, or ν_3 unresolved)symmetric stretching motion of carbon-hydrogen is dominant, and at 21 eV ν_2 and ν_1 are excited. Differential cross sections of these resonantly enhanced excitations show broad structureless bumps with respective width of 1.2 eV, 1.8 eV and 6 eV and absolute cross section of 5.10^{-17} cm^2 sr^{-1}((ν_2 at 2.6 eV), 6.10^{-18} cm^2 sr^{-1} (ν_1 at 8.2 eV) and 4.10^{-19} cm^2 sr^{-1} (ν_1 at 21 eV) for a 90 degreee scattering angle.

The symmetries of the resonant states are obtained from the angular distributions (figure 5) : at 2.6 eV the distribution is well represented by a d π wave (l=2, m=±1) giving a π_g resonant state symmetry, at 6.2 eV, the distribution is isotropic which indicates a s σ wave (l=o, m=o) and Σ_g resonant state, and at 21 eV the angular distribution is characteristic of a f σ wave (l=3, m=o) plus some contribution of a p σ wave (l=1, m=o) giving a Σ_u resonant state.

To have more informations on the role of geometry and molecular field modifications, and on the connection between shape resonances and unoccupied valence orbitals, we recently started vibrational inelastic D.C.S. in the HCN molecule which appears as intermediary between C_2H_2 and N_2, with a triple bond as N_2 and C_2H_2 and one carbon-hydrogen bond, so that the three antibonding valence orbitals $2\pi^*$ (C≡N), $6\sigma^*$(C-H) and $7\sigma^*$ (C-N) may imply the formation of three shape resonances inducing the excitation of the ν_3 (C-N stretching) and ν_2 (bending) modes at low energy, the excitation of ν_1 (C-H stretching) at intermediate energy, and excitation of ν_3 modes at high energy.

Shape resonance enhancements have been calculated to occur in the integrated elastic electron scattering cross section and in vibrational excitation cross section in CO_2 (17); they have been observed for excitation of symmetric stretch vibration via π_u, Σ_g and Σ_u resonant states with occupation of the $2\pi_u$, $5\sigma_g$ and, $4\sigma_u$ orbitals at 3.8, 10.8 and 29.5 eV respectively. (15)

The N_2O molecule is isoelectronic with CO_2 and linear in its ground state. Three unoccupied valence orbitals 3π, 8σ and 9σ have been calculated at energies of 1.9, 10.7 and 27.1 eV respectively .(24) No electron-scattering calculation has been done over a large energy range, and vibrational inelastic D.C.S has not yet been obtained for energies higher than 15 eV so that no definite assignment can be given for the shape resonances in this molecule; but recent partial photoionization cross section of the valence occupied 6σ and 7σ orbitals and angular distribution parameter measurements β from angle resolved photoelectron spectra (18) show clearly a shape resonance at a photoelectron energy of 15 eV which may imply a Σ shape resonance in inelastic electron scattering between 25 and 30 eV. The resonances observed in the excitation of the first symmetric stretch modes at 2.3 and 8.4 eV would then be π and Σ respectively which seems confirmed by the angular distributions (23). With this hypothesis the low energy shoulder observed both in inelastic D.C.S. (23) and in dissociative attachment may corresponds to the splitting of the $^2\pi$ resonant state in A' and A" states in the distorted anion.

Preliminary results for the allene molecule (CH_2=C=CH_2), isoelectronic with CO_2 and N_2O, show shape resonances with trapping of the extra electron in the 3e (π^*) orbital around 2 eV , and in the $5a_1$ (σ_g^*) orbital around 8 eV.

- Figure 4 -

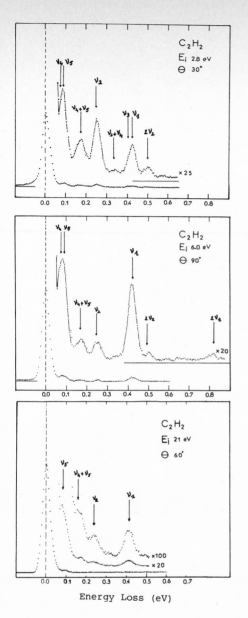

Energy loss spectra for e-C_2H_2 scattering at collision energies of 2.8 , 6.0 and 21 eV in the three resonances energy range.

The C-C symmetric stretch motion (ν_2) is dominant at 2.8 eV; the C-H stretching motion (ν_1) is dominant at 6.0 and 21 eV.

The combinations of modes ($\nu_4 + \nu_5$) ($\nu_2 + \nu_4$) are also observed.

Energy Loss (eV)

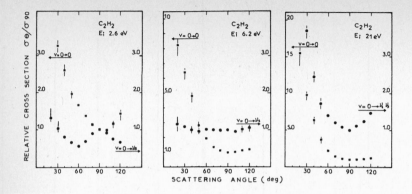

-Figure 5 -   Angular distributions for elastic and vibrational inelastic scattering in C_2H_2. The angular distributions for the ν_2 mode
at 2.6 eV, the ν_1 mode at 6.2 eV and the ν_1 mode at 21 eV agree with
Π_g, Σ_g and Σ_u resonant states respectively.

Theoretical studies of photoexcitation and ionization in H_2O[19] cast
a new light on the very broad structure extending from 2 to 15 eV in the vibrational excitation of unresolved ν_1 and ν_3 modes [21]. In photoexcitation the $4a_1$ and $2b_2$ unoccupied orbitals are observed with an energy separation of 1.9 to 2.2 eV, so that the inelastic electron scattering spectra
may result from the overlapping of two broad shape resonance states : a $^2A_1$
state centered at 8 eV and a $^2B_2$ state 2 eV higher. This hypothesis seems
confirmed by the observation of a $^2A_1$ resonant state at 2.3 eV and $^2B_2$ resonant state at 5.2 eV in H_2S (22). The two resonances being shifted lower
in energy in H_2S because of the increased molecular size and because the molecular core is more attractive. Moreover angular distribution measurements
in the broad structure in H_2O show the same evolution with energy that the
angular distribution in H_2S in the 2 to 6 eV energy range.

For these two   molecules, H_2O and H_2S, the situation is obscured
by the presence in the same energy range of Feshbach resonances known from
dissociative attachment cross section (20), which are very efficient for excitation of high vibrational levels where they can interfere with the shape
resonances.

The observation of shape resonances in polyatomic molecules and the
determination of their characteristics (energy, symmetry, active partial
waves, lifetime) is a probe of unoccupied MO's. The systematic trends observed in series of small molecules provide insights in the effect of geometrical changes and in the role of the detailed anisotropic molecular
field in the dynamics of electron-molecules processes.

REFERENCES

1) - J.L Dehmer and D.Dill in Electron and Photon Molecules Collisions
ed by T.N. Rescigno, V.Mc Koy, B.Schneider (Plenum Press N.Y.1979) p.225.
   - P.W. Langhoff same volume p.183.

2) - M.Tronc, and R.Azria Symposium on Electron-Molecule Collisions, Invited papers(Tokyo University 1979) p.105.

3) - M.Tronc, L.Malegat, and R.Azria to be published.

4) - M.R.Herman, and P.W.Langhoff Chem.Phys.Let.82, 242, 1981.

5) - W.Thiel J.Elect.Spect. and Rel.Phen. 31, 151, 1983.

6) - K.D.Jordan, and P.D.Burrow Acc. of Chem.Res. 11, 341, 1978.

7) - A.D. Mc Lean J.Chem.Phys. 32, 1595, 1960.

8) - D.W.Turner, C.Baker, A.D.Baker, and C.R.Brundle, Molecular Photoelectron Spectroscopy, Wiley N.Y. 1970.

9) - A.P.Hitchcock, C.E.Brion, J.Phys. B 14, 4399, 1981.

10) - R.Botter, W.H.Dibeler, J.A.Walker, H.M.Rosenstock, J.Chem.Phys. 44,
1271, 1966.

11) - T.Hayaishi, S.Iwata, M.Sasanuma, E.Ishiguro, Y.Morioko, Y.Iida and
Nakamura, J.Phys. B 15, 79, 1982.

12) - R.Colin, Herman, I.Kopp, Proc. 11th Astrophys. Colloq. (Université de
Liège 1977) p.355.

13) - L.E.Machado; F.P.Leal, G.Csanak, B.V.Mc.Koy and P.W. Langhoff, J.Elec.
Spect. 25, 1, 1982.

14) - A.C.Parr, D.L.Ederer, J.B.West, D.M.P.Holland, and J.L.Dehner, J.Chem.
Phys. 76, 4349, 1982.

15) - M.Tronc, R.Azria, and R.Paineau, J.Phys. Lettres 40, L323, 1979.

16) - S.F.Wong and G.J.Shulz, Phys.Rev.Let. 35, 1429, 1975.

17) - M.G.Lynch, D.Dill, J.Siegel,and J.L.Dehner, J.Chem.Phys. 71, 4249.1979.

18) - T.A.Carlson, P.R.Keller, J.W.Taylor, T.Whitley, and F.A.Grimm, J.Chem.
Phys. 79, 97, 1983.

19) - G.H.Dierchsen, W.P.Kraemer, T.N.Rescigno, C.F.Bender, B.V.Mc.Koy,
S.R.Langhoff, and P.W.Langhoff, J.Chem. Phys. 76, 1043, 1982 and references therein.

20) - D.S.Belic, M.Landau, and R.I.Hall, J.Phys. B 14, 175, 1981.

21) - G.Seng and F.Linder, B 9, 2539, 1976.

22) - K.Rohr, J.Phys. B 11, 1844, 1976.

23) - M.Tronc, L.Malegat, R.Azria, Y.Le:Coat, Abstract of contributed papers
ICPEAC XII, Gatlinburg 1981, p.372, and unpublished results.

24) - S.D.Peyerimhoff, and R.Buenker, J.Chem. Phys. 49, 2473, 1968.

LOW ENERGY EXPERIMENTS ON ELECTRON-MOLECULE
INELASTIC COLLISIONS AND THRESHOLD PEAKS:
EXPERIMENTAL ADVANCES

H. Ehrhardt, K. Jung, K.-H. Kochem, and W. Sohn
Fachbereich Physik der Universität Kaiserslautern
D-6750 Kaiserslautern, West Germany

1. Introduction

A few years ago Linder and Rohr (Linder and Rohr 1976) have observed that the cross
sections for vibrational excitation of dipole molecules may be very large close to
threshold. Also non-dipole molecules as CH_4 and SF_6 show such threshold structures
(Rohr 1979, 1980). Several theoreticians have discussed this effect in model and ab
initio calculations (Dubê and Herzenberg 1977, Nesbet 1980, Morrison 1982, Morrison
and Lane 1979, Domcke and Cederbaum 1980, Gianturco and Thompson 1976, etc.). The
most important result of these theoretical papers was that a virtual state could produce the shape and the magnitude of such a threshold peak. The purpose of the present
paper is to continue the experiments of Linder and Rohr in order to obtain a more general view of vibrational threshold excitation especially with respect to different
interaction potentials (dipole, quadrupole, polarizability).

2. Apparatus

The electron spectrometer (Sohn et al. 1982) consists of an electron gun, which is
differentially pumped and an electron detector, both equipped with a cylindrical
127°-monochromator. The detector is rotatable between - 25° to 110°. Both systems as
well as the gas inlet tube are usually heated during scattering operation. The typical temperature is 200°C.

3. Experimental Procedures

In order to get reliable results in the near threshold region down to collision
energies of 50 to 100 meV the measurements have to be performed under the following
conditions:

1st: the overall energy resolution should be at least 25 meV in order to resolve vibrational modes and narrow threshold structures.

2nd: the optical systems must produce a well-defined electron beam of very low primary energies. This can be achieved by an active potential supply for both lense systems (gun and detector) and by controlling the tuning of the electron optics via calculated electron trajectories.

3rd: the transmissions of the electron optics have to be nearly constant; in any case they have to be known exactly. The transmission behaviour is controlled by comparing the measured shapes of the elastic e^-+He energy and angular dependencies with accurate theoretical cross sections (Nesbet 1980).

Absolute cross sections are determined using a flux-rate method (Sohn et al. 1982).

4. Results

1. OCS

The OCS molecule has a permanent dipole moment of 0.715 Debye. Therefore, one expects a strong increase of the elastic cross section towards low collision energies as well as a strong forward peaking of the elastic differential cross section. This has been predicted by Itikawa (Itikawa 1969), and was found experimentally by Linder and Rohr for the case of HCl (Linder and Rohr 1976). We have found the same behaviour for OCS

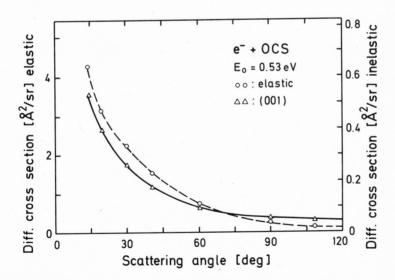

Fig. 1: Angular dependencies of the DCS for elastic scattering and the excitation of the asymmetric stretch mode (001) of OCS at a collision energy of E_0=0.53 eV.

for primary energies down to 0.3 eV and scattering angles from 110° to 15°. For example, the angular dependencies of the elastic scattering and the excitation of the asymmetric stretch mode (0 0 1) at E_o = 0.53 eV is shown in figure 1.

Figure 2 shows an example of the energy dependence of the excitation of the asymmetric stretch vibrational mode for a scattering angle of θ = 20°.

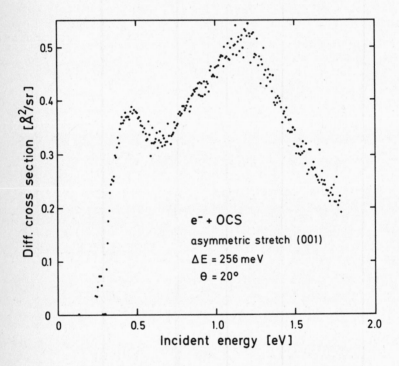

Fig. 2: Energy dependence of the DCS for the excitation of the asymmetric stretch mode (0 0 1) of OCS at a scattering angle of θ = 20°.

The differential cross section shows a broad peak centered around 1.2 eV which is due to a Π_u-shape-resonance (Lynch and Dill 1979). Close to threshold (E_{th} = 256 meV) another structure is visible. Our measurements show that the width of this threshold peak increases with decreasing scattering angle whereas the energetic position of the maximum shifts towards higher energies. This is in good agreement with the qualitative behaviour one would obtain for pure direct excitation via a dipole potential in Born approximation, assuming realistic derivatives of the dipole moment with the internuclear distance.

The bending vibrational modes (0 1 0) and (0 2 0) show qualitatively the same energy and angular behaviour. They also have similar absolute values of the cross sections. The symmetric stretch vibrational mode has much lower cross sections and could not be detected at any scattering angle and collision energy.

2. CO_2

The electron scattering by CO_2 molecules has been investigated by different authors experimentally as well as theoretically (see for example Ferch et al. 1981, Morrison and Lane 1979, Morrison 1982). A $^2\Pi_u$-shape-resonance around 3.8 eV and a virtual state with a corresponding scattering length of about -6 a_0 are well established (Ferch et al. 1981); the expected energy dependence of the elastic differential cross section at a scattering angle of θ = 90° can be seen in figure 3.

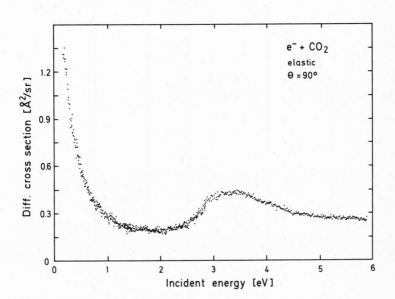

Fig. 3: Energy dependence of the elastic DCS for CO_2 at θ = 90°. The enhancement to-wards lower energies is due to a virtual state. The broad bump around 3.8 eV is caused by a $^2\Pi_u$-shape-resonance.

From theory one would expect that the virtual state dominates the vibrational excitations near the corresponding thresholds (Whitten and Lane 1982). The surprising result of our measurement (see figure 4) is the fact that the contribution of the virtual state to the vibrational excitation seems to be small or at least much smaller as expected (figure 5).

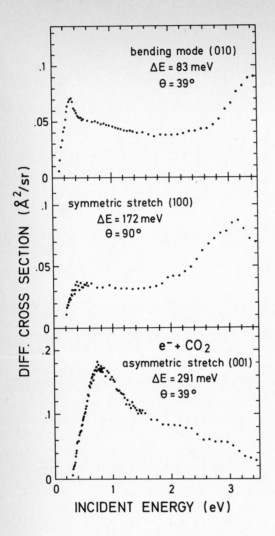

Fig. 4:

Energy dependencies of the DCS for the excitation of the three fundamental vibrational modes of CO_2 at $\theta = 39°$ for the infrared active modes and $\theta = 90°$ for the raman active mode (1 0 0).

This is partly deduced from the angular dependencies of the different vibrational modes. The infrared active modes (bending and asymmetric stretch mode) show, even at energies close to threshold, a strong forward peaking, which is in contradiction to the s-wave angular behaviour of a virtual state. A first estimation yields that similar to the corresponding OCS data, the shapes and the magnitudes of the (0 1 0) and (0 0 1) vibrational excitations can be explained by direct processes (Domcke and Estrada, private communication).

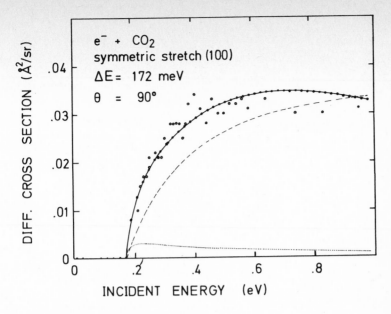

Fig. 5: Energy dependence of the DCS at θ = 90° for the excitation of the symmetric
stretch mode (1 0 0) of CO_2. The circles are our experimental result. The
full curve represents a theoretical calculation by Domcke and Estrada including parts of the DCS due to the virtual state (dotted line) and due to
background scattering (dashed line).

Although the angular dependence of the infrared inactive mode (symmetric stretch mode
(1 0 0), ΔE = 172 meV) is isotropic, which would be consistent with the virtual state
concept, the expected large threshold peak (Whitten and Lane 1982) could experimentally not be found (see figures 4 and 5), so that even for this mode the above argument holds.

We also performed very low electron energy measurements on CH_4, which has no permanent
dipole nor quadrupole moment and therefore has other long range interactions as OCS
and CO_2. The results of these investigations have been published (Sohn et al. 1982).

5. Conclusions

The threshold excitation of vibrations has been determined for target systems with
different interaction potentials. The main result of our investigation is that non-
resonant direct excitation processes play an important role in the threshold energy
region. Infrared active vibrational modes are excited through electron impact showing
a threshold peak and a strong forward scattering, whereas Raman active modes are ex-

cited with lower intensities and less pronounced threshold peaks and relatively flat angular distributions. This seems to show that non-adiabatic effects play an important role.

References

Domcke W and Cederbaum LS 1980 J. Phys. B 14 149

Dubé L and Herzenberg A 1977 Phys. Rev. Lett. 38 820

Ferch J, Masche C and Raith W 1981 J. Phys. B 14, L97

Gianturco FA and Thompson DG 1976 J. Phys. B 9 383

Itikawa Y 1969 J. Phys. Soc. Japan 27 444

Lane NF 1980 Rev. Mod. Phys. 52 29

Lynch MG and Dill D 1979 J. Chem. Phys. 71 4249

Morrison MA and Lane NF 1979 Chem. Phys. Lett. 66 527

Nesbet RK 1980 J. Phys. B 13 L193

Rohr K 1979 "Symposium on Electron-Molecule-Collisions Invited Papers" 67

Rohr K 1980 J. Phys. B 13 4897

Rohr K and Linder F 1976 J. Phys. B 9 2521

Sohn W, Jung K and Ehrhardt H 1982 J. Phys. B 16 891

Whitten L and Lane NF 1982 Phys. Rev. A 26 3170

DISSOCIATIVE ATTACHMENT IN POLAR

AND HIGHLY POLAR MOLECULES

J.P. ZIESEL
Laboratoire des Collisions Atomiques et Moléculaires[+], Université Paris-Sud
Bât. 351, 91405 ORSAY Cedex, France

The behavior of electron scattering by hydrogen halides HX (dipole moment 0.4-1.8 D) has arisen the interest to study the collisions of electrons with highly polar molecules, like the alkali halides MX. New phenomena have been observed in hydrogen halies, stimulating the calculation of electron scattering with polar and highly polar molecules, recently reviewed[1]. Concerning the dissociative attachment in hydrogen halides, the most abundant ion is X^- ; striking features observed in the X^- formation are the coupling of the dissociation and autodetachment continua[2,3] and the large effect of the initial rovibrational energy of the HX ground state on the absolute cross-section[4-7]. The H^- ion (about 1 % of X^-) is formed thru 3 dissociative resonant states, with the $^2\Sigma_{1/2}$ and $^2\Pi_{3/2}$ HX$^-$ states correlated to the dissociation asymptote $H^- + Br(^2P_{3/2})$ and the $^2\Pi_{1/2}$ state correlated to $H^- + Br(^2P_{1/2})$. Coupling between the $^2\Sigma_{1/2}$ and $^2\Pi_{1/2}$ states has been observed in HBr[8] and in HCl[9,10]. This effect could be attributed to non-adiabatic coupling between these states with the same $\Omega = 1/2$ symmetry[8] and/or to electronic coupling via the non-resonant e-HX continuum[11].

The alkali halides have a very large dipole moment (6-12 D) ; the electron-dipole interaction is then strong enough to bind an electron in the anion ground state MX$^-(1^2\Sigma)$[12] and possibly in some excited states[13-16]. The first evidence of a dipole bound state came from a photodetachment experiment[17] in LiCl ; the experimental (0,61 eV) and calculated[13] (0.54 eV) binding energies are in good agreement. Dissociative electron attachment will only probe some excited resonant state in the electron scattering continuum ; measurements of the cross-section dependence over the energy and angle and of the kinetic energy of the product negative-ion will help to determine the potential energy curves of these states and their symmetry.

Experimental

Dissociative attachment in sodium halides is studied with two different crossed-beam mass spectrometers. In the magnetic apparatus, the electron beam is monochromatized thru a trochoïdal monochromator and crosses the molecular beam effusing from a Knudsen oven ; the negative ions are extracted at around 90° of the two beams and mass-analyzed with a magnetic sector. In the electrostatic apparatus[18], the electron beam from a 127° cylindrical filter crosses the molecular beam from a two-stage oven with a 14 x 1 mm dia. exit channel. The negative ions are energy- analyzed in a second 127° electrostatic filter, which can rotate from 30 to 145 ° with respect to the electron beam ; the ions are then analyzed with a quadrupole mass-filter.

The dissociative attachment (DA) process studied is $e(E_i)$ + NaX → $Na^-(E_R, \theta)$ + X, where E_i is the incident electron energy, E_R the kinetic energy of the Na^- fragment and θ the angle between the dissociation axis and the electron beam. The magnetic mass spectrometer is used to measure the energy dependence of the DA cross-section and to calibrate the electron energy scale with the O^-/CO vertical onset at 9.62 eV. The elec trostatic mass spectrometer can be operated in different modes. A <u>kinetic energy E_R spectrum</u>, at fixed E_i and θ, gives identification of the dissociation limit and thus of the fragment states ; this is particularly important in the alkali halides, where DA can occur from the dimer as well as from the monomer. <u>An ion-yield spectrum</u>, at fixed θ, gives the yield of ions arising from a given DA process as a function of electron energy E_i. <u>An ion angular distribution</u>, at fixed E_i and E_R, gives the depen- dence over θ,that can be compared to theory for the determination of the resonant state symmetry.

<u>Experimental results</u> : <u>dissociative attachment in NaBr</u>

Although work has been completed in NaX (X = F, Cl, Br) and partial results obtained in NaI, LiF and LiCl[19], discussion will be limited to DA in NaBr. The Na^- ion-yield spectra for the DA processes correlated to the dissociation into Na^- + Br ^2P are shown in figure 1.

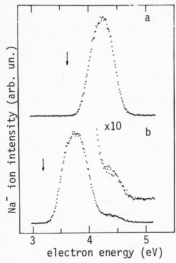

Fig. 1. Na^- ion-yields in NaBr.
  a) dissociation to Br $^2P_{1/2}$;
  b) dissociation to Br $^2P_{3/2}$.
The arrows indicate the position of the dissociation thresholds.

There is one process leading to Br $^2P_{1/2}$, unlike DA in HBr where 2 processes are ob- served[18], one of them arising from the coupling of the $^2\Sigma_{1/2}$ state with the $^2\Pi_{1/2}$ state ; two processes are observed in the Br $^2P_{3/2}$ dissociation channel. The peak energies are quite different, suggesting that 3 resonant states are involved. However, as discussed below, the attribution of the second peak (Br $^2P_{3/2}$ dissocia- tion) to a DA process in the monomer is doubtful As seen on the figure, the onsets of the two main processes are very close to the dissociation energies ; the potential energy curves should then be flat at large R or even have a shallow minimum.

The angular dependencies of the Na^- ion formation, normalized at the angle of 90°, are shown at four different energies (fig. 2). These angular distributions are proportional to the DA differential cross-section $d\sigma/d\Omega$.

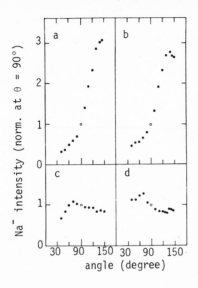

Fig. 2. Experimental angular distributions of Na^- from NaBr : dissociation to $Br\ ^2P_{3/2}$ at
a) $E_i = 3.77$ eV ; b) $E_i = 3.97$eV; dissociation to $Br\ ^2P_{1/2}$ at
c) $E_i = 4.27$eV and d) $E_i = 4.57$eV.

The experimental shapes are quite different for the processes leading to $Br\ ^2P_{3/2}$ and $Br\ ^2P_{1/2}$, meaning that indeed states with different symmetries are involved. A common behavior is the strong asymmetry of the curves with respect to $\theta = 90°$;this is also observed in Na^-/NaF and Na^-/ NaCl. An asymmetrical behavior in the differential cross-section has been already reported in the O^-/CO process at 9.62 eV[20] ; Hall et al. have successfully analysed the data by the mixing of two partial waves in the electron attachment on the heteronuclear diatomic molecule CO, using the theory of O'Malley and Taylor[21].

However, by looking to the hydrogen halides and sodium halides series (table below), the asymmetry sems in some way related to the dipole moment of the neutral molecule.

If we put aside the H^-/HI case, where the asymmetry can be explained[22] like in O^-/CO, by the mixing of two partial waves, one can see that the asymmetry doesn't exist in HBr and HCl, appears in HF and becomes more pronounced in highly polar molecules. It should be noted that a slight asymmetry has also been observed in the H^- differential cross-section in the triatomic polar molecules H_2O($\mu = 1.85$ D)[25] and H_2S ($\mu = 0.97$ D)[26] ; it could also well be related to the role of the electron-dipole interaction in the attachment process. The theory of O'Malley and Taylor[21] does not take it into account, as it is assumed that the e-molecule interaction is spherically symmetric at large distance r_e. Teillet-Billy and Gauyacq[27,28] have recently introduced the dipolar interaction and then could generalize the O'Malley and Taylor theory to polar and highly polar molecules.

Table I

	μ	Symmetry/angle $\Theta = 90°$				Ref.
H^-/HI	0.44D		no		no	22
H^-/HBr	0.8 D	$^2\Sigma_{1/2}$ yes		$^2\Pi_{3/2,1/2}$ yes		18
H^-/HCl	1.08D	$^2\Sigma$ yes		$^2\Pi$	yes	8
H^-/HF	1.8 D		no		no	9,23
Na^-/NaF	8.1 D		no		no	
$Na^-/NaCl$	9 D		no		no	24
$Na^-/NaBr$	9.1 D		no		no	

DA differential cross-section with a dipolar field : theory of Teillet-Billy and Gauyacq.

Only a brief outline will be given, as the theory will be exposed in detail in a forthcoming paper[28]

i) angular eigenmodes of the e-molecule scattering. In the wave equation $(H_e-E)\ F = \mathbf{O}$ the electronic hamiltonian contains the dipolar interaction, proportional to $\mu\cos\theta$ The eigenfunctions F can then be expressed as a product $\psi_e(r_e)\phi_e(\hat{r}_e)$, where r_e are the radial coordinates and r_e the polar angles of the incident electron. The wave equation is then separable into a radial and an angular equations[29]. When the dipole moment μ is equal to zero, the angular part of the eigenfunctions are spherical harmonics. When μ is different from zero, the angular eigen functions $\phi_n^m(\hat{r}_e)$ can conveniently be expressed as a sum of spherical harmonics $\phi_n^m(\hat{r}_e) = \sum\limits_{\ell=m}^{\infty} A_\ell^m\ Y_\ell^m(\Theta,\varphi)$ [1] The coefficients A_ℓ^m are dependent of μ.

ii) electron wave function $\psi_e(\vec{r}_e,\ \vec{k})$

The angular eigen modes ϕ_n^m are a complete set for angular functions; the expression of ψ_e is $\sum\limits_{n,m}\ \psi_{n,m}(kr_e)\phi_n^{m*}(\hat{k})\ \phi_n^m(\hat{r}_e)$ [2], where the role of the electron momentum and electron coordinates are symmetrical.

iii) application of the treatment of O'Malley and Taylor.

In this theory, the resonant state is coupled to the initial state by the electronic matrix element $V(\vec{R})$, expressed in the molecular frame as

$$V(\vec{R}) = \psi_r(\vec{r}_{T+e},R)\ H_1\ \psi_T(\vec{r}_T,R)\psi_e(\vec{r}_e,\vec{k})d\vec{r}_{T+e} \qquad |3|$$

where ψ_r is the resonant state electronic wave function. V depends on the orientation of the momentum $\vec{k}$ with respect to the internuclear axis through the initial state wave function. When ψ_e from (2) is introduced into the expression (3), the coupling matrix element is now expressed as a sum of angular eigenmodes.

$$V \sim \sum_{n=m}^{\infty} v_n^m \phi_n^m \, (\hat{k}) \qquad [4]$$

m is given by $\Lambda_r - \Lambda_t$, where Λ_r and Λ_t are the axial angular momenta of the resonant state and the target state respectively. In the slow rotation approximation, the molecular axis is assumed not to rotate during the collision and so the dissociation will proceed along the initial direction of the molécular axis. Finally, the differential cross-section is dependent over the polar angles $\hat{R}$ -with respect to the electron momentum direction- only through ϕ_n^m and takes the form (Teillet-Billy and Gauyacq)

$$d\sigma/d\Omega \sim \left| \sum_{n=m}^{\infty} a_n^m \phi_n^{m=\Lambda_r - \Lambda_t} (\theta,\varphi) \right|^2 \qquad [5]$$

As an example, the relative DCS, normalized for $\theta = 90°$, are shown(figure 3) for an initial molecular state Σ ($\Lambda_t = 0$) and a resonant state Π ($\Lambda_r = 1$), assuming that only the angular mode ϕ_1^1 is active.

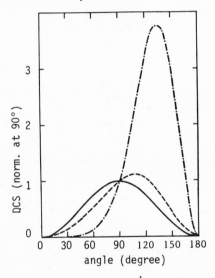

Fig. 3. Relative DCS for a $\Sigma \to \Pi$ attachment transition

— $\mu = 0$ $\qquad \phi_1^1 = Y_1^1$;

-- $\mu = 1.8$ D $\qquad \phi_1^1 = 0.995 \, Y_1^1 -$

$- 0.096 \, Y_2^1 + 0.004 \, Y_3^1$;

—· $\mu = 9$ D $\qquad \phi_1^1 = 0.834 \, Y_1^1$

$- 0.528 \, Y_2^1 + 0.157 \, Y_3^1$

$- 0.028 \, Y_4^1 + 0.003 \, Y_5^1.$

When there is no dipole moment, ϕ_1^1 is reduced to the spherical harmonic Y_1^1 ; when the dipole moment is increased, the DCS becomes more and more asymmetrical. For $\mu=1.8$ D, the agreement with the experimental data on $\bar{H}/HF$ at 9.5 eV is very good[27,28] ; in this case, two spherical harmonics only are involved in the process and a fit was obtained using the O'Malley and Taylor theory[30]. However this spherical harmonics combination comes naturally within the Teillet-Billy and Gauyacq treatment ; furthermore the number of spherical harmonics, even with only one active angular mode, increases rapidly when dealing with highly polar molecules (figure 3) and a spherical harmonics fit would become impracticable.

Discussion : <u>symmetry of NaBr$^-$ dissociative states</u>

Experimental angular distributions can now be compared with the theoretical DCS to
determine the symmetry of the resonant states. As the target is in the ground state
X $^1\Sigma$ ($\Lambda_t = 0$), the eigenmodes to be considered will be $\phi_{n \geqslant o}^o$, for a resonant state Σ,
and $\phi_{n \geqslant 1}^1$, for a resonant state Π.

The angular dependence for the dissociative attachment process at 3.77 eV leading to
Na$^-$ + Br $^2P_{3/2}$ formation is shown in fig. 4

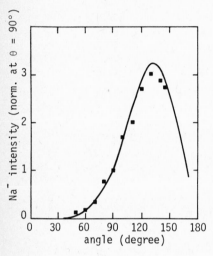

Fig. 4. Angular distribution of
$\quad$ Na$^-$(+ Br $^2P_{3/2}$)
■ Experiment E_i = 3.77 eV (corrected)
— $|\phi_1^1|^2$ corrected for NaBr thermal
$\quad$ motion (1000 K).

Both the experiment and the theoretical DCS, proportional to $|\phi_1^1|^2$ have been corrected.
The agreement is quite good as well as at a higher energy (Fig. 4b) ; one can then
attribute this process to the attachment in the resonant state NaBr$^-$ $^2\Pi_{3/2}$.

The observed DCS is corrected by sin θ, assuming the worst case for the variation of
the collision volume with the angle of observation. The other effect which should be
corrected is due to the thermal motion of the target. Besides the Doppler broadening
of the fragment ion kinetic energy distribution, the thermal translational motion can
induce a loss of anisotropy ; this small effect has been discussed in detail for the
H$^-$/H$_2$ DA processes by Fiquet-Fayard et al[31] . The theoretical angular distribution can
be corrected at every angle θ using the formulation of Misakian et al[32] ; these corrected theoretical DCS, assuming a transverse translational temperature of 1000 K,
are shown in fig. 4-6 and compared to the experiments, corrected by sin θ. No attempt

has been made to include the loss of anisotropy due to the rotational motion ; this contribution is estimated to be of about the same importance as that of the translational motion[28].

The experimental angular distributions for the (Na$^-$ + Br $^2P_{1/2}$) process are peaked at about 70° and then decrease slowly (Fig. 2c, 2d), more steeply when the sin θ correction is applied (Fig. 5)

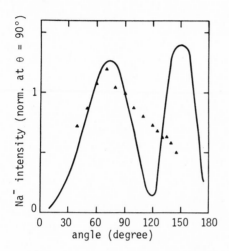

Fig. 5. Angular distribution of
Na$^-$(+Br $^2P_{1/2}$)

▲ Experiment E_i = 4.57 eV (corrected):

— $|\phi_2^1|^2$ corrected for NaBr thermal motion (1000 K)

The only angular mode which give a DCS with a maximum around 70° is the ϕ_2^1 mode. However, it is clearly apparent on Fig. 5 that the theoretical DCS cannot be fitted to expériment at large angles. Unlike in the Na$^-$/NaCl case[24], more than one angular mode are thus needed and a sum $|a_2\phi_2^1 + a_1 \phi_1^1|^2$ could well be more realistic so as, in particular, to reduce the gap between experiment and $|\phi_2^1|^2$ around 120°. This would then be in agreement with a resonant state $^2\Pi_{1/2}$ correlated to the Na$^-$ + Br $^2P_{1/2}$ asymptote. However, one cannot rule the possibility that the theory is no more strictly valid for molecules like NaBr, where the spin-orbit interaction becomes important and Λ is no more a good quantum number.

There is a $^2\Sigma_{1/2}$ state correlated to Na$^-$ + Br $^2P_{3/2}$ formation. The angular distribution close to the onset of the ion-yield spectrum are much higher than the corrected $|\phi_1^1|^2$ (Fig. 6).

The sin θ correction on the experimental data at E_i = 3.58 eV is the most severe that could be applied. The experiment agrees with the corrected $|\phi_0^0|^2$ until 120°, giving credence to a state of Σ symmetry. The shape over 120° show that there is more than one angular mode involved ; the DCS could be a sum $(a_0\phi_0^0)^2 + (a_1\phi_1^1)^2$ as the smooth ion-yield spectrum (fig. 1b) shows that the $^2\Pi_{3/2}$ state is still contributing at these lower energies. So the low intensity second peak at E_i = 4.39 eV would not be dissociation from the resonant state $^2\Sigma_{1/2}$ but would arise from a dissociative attachment process in the dimer $(NaBr)_2$.

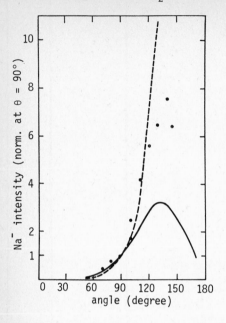

Fig. 6. Angular distribution of
$Na^-(+Br\ ^2P_{3/2})$

•   Experiment E_i=3.58 eV (corrected);

— $|\phi_1^1|^2$ corrected for NaBr thermal motion (1000 K)

Conclusion

The role of the dipolar interaction in electron-polar molecules collisions has been made in evidence in our dissociative attachment experiments. The differential cross-sections in the sodium halides exhibit a strong asymmetry which must be characteristic of molecules with a large dipole moment. This asymmetry is clearly explained when the dipolar interaction is introduced in an extended treatment of the O'Malley and Taylor theory. Although the interpretation of the experimental data is straight forward in HF, NaF and NaCl, the situation is not so clear in NaBr, stressing the importance of next studies in NaI where spin-orbit interaction is still more important.

Acknowledgments

A large credit should be given for the experimental work to Roger AZRIA, Dominique TEILLET-BILLY, Robert ABOUAF, Philippe GIRARD and Lucien BOUBY. Also we are grateful to Jean-Pierre GAUYACQ and Dominique TEILLET-BILLY for the mutual interaction between experiment and theory.

[+]Laboratoire associé au C.N.R.S.

References

1. N.F. Lane, Rev. Mod. Phys. 52 29 (1980) ; D.W. Norcross and L.A. Collins, in Advances in Atomic and Molecular Physics 18 341 (1982).

2. J.P. Ziesel, I. Nenner and G.J. Schulz, J. Chem. Phys 63 1943 (1975).

3. F. Fiquet-Fayard, J. Phys. B 7 810 (1974).

4. M. Allan and S.F. Wong, J. Chem. Phys. 74 1687 (1981).

5. A. Herzenberg, Notas de Fisica 5 225 (1982).

6. J.N. Bardsley and J.M. Wadehra, J. Chem. Phys. 78 7227 (1983).

7. D. Teillet-Billy and J.P. Gauyacq, Lecture Notes in Chemistry, this volume.

8. R. Azria, Y. Le Coat, D. Simon and M. Tronc, J. Phys. B 13 1909 (1980.

9. R. Azria, in Physics of Electronic and Atomic Collisions (S. Datz ed., North Holland, Amsterdam, 1982).

10. A. Huetz and J. Mazeau, private communication.

11. A.U. Hazi, J. Phys. B 16 L29 (1983) ; Lecture Notes in Chemistry, this volume.

12. K.D. Jordan, Acc. Chem. Res. 12 36 (1979).

13. K.D. Jordan and W. Luken, J. Chem. Phys. 64 2760 (1976)

14. A.U. Hazi, J. Chem. Phys. 75 4586 (1981).

15. W.R. Garrett, J. Chem. Phys. 77 3666 (1982).

16. I.I. Fabrikant, J. Phys. B 16 1269 (1983).

17. J.L. Carlsten, J.R. Peterson and W.C. Lineberger, Chem. Phys. Lett. 37 5 (1976).

18. Y. Le Coat, R. Azria and M. Tronc, J. Phys. B 15 1569 (1982).

19. D. Teillet-Billy, L. Bouby and J.P. Ziesel, 12th ICPEAC, abstracts p. 407 (1981).

20. R.I. Hall, I. Cadez, C. Schermann and M. Tronc, Phys. Rev. A 15 599 (1977).

21. T.F. O'Malley and H.S. Taylor, Phys. Rev. 176 207 (1968).

22. Y. Le Coat, R. Azria and M. Tronc, to be published.

23. R. Abouaf, R. Azria and D. Teillet-Billy, to be published.

24. J.P. Ziesel, R. Azria and D. Teillet-Billy, to be published.

25. R.I. Hall and S. Trajmar, J. Phys. B 7 L458 (1974).

26. R. Azria, Y. Le Coat, G. Lefèvre and D. Simon, J. Phys. B 12 679 (1979).

27. D. Teillet-Billy and J.P. Gauyacq, 13th ICPEAC, Abstracts p. 297, Berlin (1983).

28. D. Teillet-Billy and J.P. Gauyacq, submitted to J. Phys. B.

29. M.H. Mittleman and R.E. von Holt, Phys. Rev. 140 A726 (1965).

30. D. Teillet-Billy, private communication.

31. M. Tronc, F. Fiquet-Fayard, C. Schermann and R.I. Hall, J. Phys. B 10 305 (1977).

32. M. Misakian, J.C. Pearl and M.J. Mumma, J. Chem. Phys. 57 1891 (1972).

PART II

LOW-ENERGY ELECTRON-MOLECULE
SCATTERING: THE THEORETICAL
TOOLS

R-MATRIX THEORY OF ELECTRON-MOLECULE COLLISIONS

C.J. Noble[+], P.G. Burke[*] and S. Salvini[*]

[+]SERC Daresbury Laboratory, Daresbury, Warrington WA4 4AD, U.K.
[*]Department of Applied Mathematics and Theoretical Physics, The Queen's University of Belfast, Belfast BT7 1NN, Northern Ireland

Over the past few years we have been developing a general computer program to enable electron collisions with diatomic molecules to be calculated within the framework of R-matrix theory. The modifications of the basic R-matrix formalism which we have adopted in order to treat electron-molecule collisions have been described in some detail in recent reviews by Buckley et al [1] and by Burke and Noble [2]. However there has been considerable progress during the past year so we will present here an outline of some of the computational improvements which have been carried out together with some of the results which have been obtained during this period.

Briefly, the theory assumes that the total electronic molecular wave function for each internuclear separation may be expanded within a finite spherical region of configuration space in the form [3]

$$\Psi(x_1 \ldots x_{N+1}) = A \sum_i \Phi_i(x_1 \ldots x_N) F_i(x_{N+1}) + \sum_i \phi_i(x_1 \ldots x_{N+1}) b_i \qquad 1$$

where x_i denotes the space-spin coordinates of the i-th electron, Φ_i are target wave functions, F_i describes the motion of the scattered electron and ϕ_i are quadratically integrable correlation functions. The Φ_i and ϕ_i are represented by STO's centred on the target nuclei, while F_i is expanded in terms of STO's centred on the nuclei and numerically defined basis functions on the centre of gravity. These basis functions are constructed so that only those on the centre of gravity remain non-negligible on the boundary of the R-matrix internal region and so provide the connection to the single-centre expansion which is used to solve the spherically symmetric scattering problem in the external region. A Bloch operator must be added to the Hamiltonian in the inner region to form a Hermitian operator. The matrix representation of this latter operator is then constructed and diagonalised using a modified version of the ALCHEMY molecular structure package [4,5]. In the outer region the resultant coupled differential equations are solved using an R-matrix propagation code [6] and an asymptotic Gailitis expansion [7,8].

The construction of the scattering orbitals, F_i, is a crucial factor in obtaining an accurate representation of the system wavefunction for scattering energies spanning the low and intermediate energy ranges. We obtain numerical basis functions by solving the model single-channel scattering problem

$$\left[\frac{d^2}{dr^2} - \frac{\ell_i(\ell_i + 1)}{r^2} + V_o(r) + k_j^2 \right] u_{ij}(r) = 0 \qquad 2$$

(V_o is the static potential of the target molecule) subject to the requirement that the logarithmic derivative of the solutions vanishes on the R-matrix boundary. The set of functions which is generated is complete over a known energy range. Continuum orbitals are then determined by taking linear combinations of the occupied target orbitals and the numerical basis functions using a procedure similar to Lagrange orthogonalisation which will be described elsewhere [9]. An orthonormal set of orbitals is obtained which is again complete over a known energy range.

Recently results have been obtained for electron scattering by H_2, N_2, F_2, CO, LiH and H_2^+ molecules. Two models have been investigated. In the static exchange

(SE) model the correlation functions ϕ_i are formed by taking the SCF target wave function and allowing the scattering electron to occupy one of the virtual orbitals of the correct symmetry. In the SEP model account is taken of charge polarization effects by in addition allowing single particle excitations of one of the target valence electrons so that correlation functions ϕ_i with a two particle - one hole (2p-1h) character are formed.

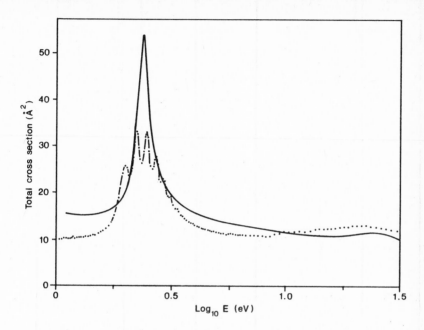

Figure 1. Total cross sections in $Å^2$ for e-N$_2$ scattering as a function of the incident energy in eV. Solid line, SEP model R-matrix results [11]; Dotted line, experimental measurements of Kennerly [10].

The ability of the SEP model to accurately reproduce experimental results is illustrated in Figure 1 where we compare our results for the total cross section for e-N$_2$ scattering with the absolute experimental measurements of Kennerly [10]. The discrepancy between theory and experiment for scattering energies up to 35 eV is generally small. Even the over-estimate by theory at the lowest energies can most probably be accounted for by defects in the target representation including the neglect of target correlation [11]. The failure to account for the detailed structure in the region of $^2\Pi_g$ resonance is of course a consequence of the fact that our calculations are at a fixed internuclear separation and neglect the coupling to the nuclear motion. The inclusion of effects arising from the nuclear motion is currently being investigated.

One of the special difficulties for ab initio calculations presented by electron scattering by polar molecules is the large number of partial waves which must be retained in order to obtain converged results. The multicentre expansion of the wave function (equation 1) implicitly includes contributions from high partial waves and so minimises this problem allowing us to obtain converged results for both the weakly polar CO and strongly polar LiH molecules. For example, in Figure 2 we show some results we have obtained for the eigenphase sum for e-CO scattering in the $^2\Pi$ symmetry [12]. The position and width of the resonance obtained by fitting the eigenphase sum for the SE model using the McLean and Yoshimine target wave function [13] are 3.49 eV and 1.97 eV, respectively. These agree well with the corresponding

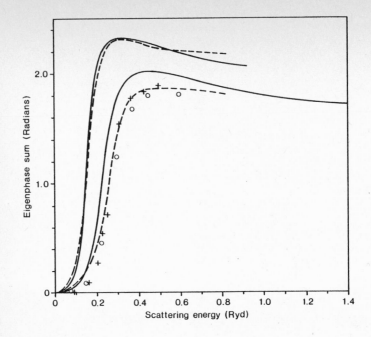

Figure 2. Eigenphase sum in radians for e-CO scattering in the $^2\Pi$ symmetry as a function of the scattering energy in Rydbergs. R-matrix results [12] using a McLean-Yoshimine wave function [13]: lower dashed curve, SE model, upper dashed curve, SEP model; using a Nesbet wave function [14]: lower solid curve, SE model, upper solid curve, SEP model. Open circles, Collins et al [15]; open squares, Levin et al [16].

values of 3.54 eV and 2.07 eV obtained by Collins et al [15]. The inclusion of polarization effects by means of the SEP model shifts the resonance to lower energies and reduces its width. For example using a Nesbet target wave function [14] the resonance occurs in the SEP model at an energy of 1.72 eV and with a width of 0.75eV.

We have begun to study electronic excitation processes by considering low energy electron impact on the hydrogen molecular ion using SE and SEP models in which both the ground $X^2\Sigma_g^+$ and first excited $^2\Sigma_u^+$ states of H_2^+ are retained in the sum over target states, Φ_i, in equation 1. The numerical accuracy of our SE model results may be verified by comparing with equivalent results obtained by Collins and Schneider [17] using the linear algebraic equations method. However, our SEP model results are the first using an ab initio method in which both channel coupling and polarization effects are explicitly represented [18]. The most distinctive feature of the results is the appearance of an infinite series of Rydberg resonances converging toward the threshold of the excited state. The first two members of such a sequence may be seen in Figure 3 where we show eigenphase sums for the $^3\Pi_g$ symmetry as a function of the scattering energy. For this symmetry even the lower members of the Rydberg series are very narrow and the effect of polarization is slight. More pronounced effects are found for the lower energy resonances which occur for scattering in the $^1\Sigma_g^+$ and $^1\Sigma_u^+$ symmetries [18,19]. Some of the positions and widths we have obtained for these resonances are shown in Table I.

The results obtained to date, using the R-matrix method permit a number of conclusions to be drawn. First, when combined with the use of numerical continuum orbi-

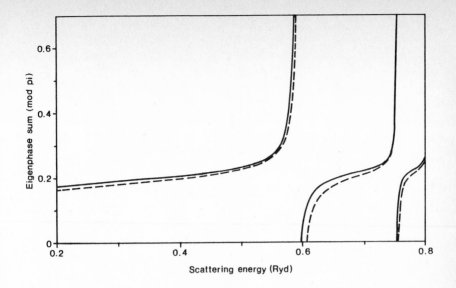

Figure 3. Eigenphase sum for e-H$_2^+$ scattering in the $^3\Pi_g$ symmetry as a function of the scattering energy in Rydbergs. R-matrix results [19], solid curve, SEP model; dashed curve, SE model.

Table I

Position and width of the two lowest e-H$_2^+$ resonances in the $^1\Sigma_g^+$ symmetry. E_r is the resonance energy, Γ is the width.

	E_r(Ryd)	Γ(Ryd)	
1	.4440	.095	R-matrix, SE model [19]
	.4088	.106	R-matrix, SEP model [19]
	.4420	.104	Collins and Schneider, two state model [17]
	.395	.103	Schneider and Collins, optical potential 2 [18]
2	.7355	.010	R-matrix, SE model [19]
	.7321	.010	R-matrix, SEP model [19]
	.729	.011	Schneider and Collins, optical potential 2 [18]

tals the method is numerically stable and the computational expense is competitive with alternative methods. Second, as mentioned at the outset, the results have been obtained with a general computer code so the application of the method to a wide range of diatomic targets and scattering processes is now feasible. The extension of the method to describe photoionization and dissociative attachment and recombination processes is also in progress.

Acknowledgements

We would like to thank our colleagues J. Tennyson and U. Lammana for permission to quote unpublished results.

References

1. Buckley, B.D., Burke, P.G. and Noble, C.J., 1983, to be published in "Electron Molecule Collisions" Ed. I. Shimamura and K. Takayanagi. (Plenum Press, New York).

2. Burke, P.G. and Noble, C.J., 1983, Comments At. Mol. Phys., 12, 301.

3. Burke, P.G., Mackey, I. and Shimamura, I., 1977, J. Phys. B, 10, 2497.

4. McLean, A.D., 1971, Proc. Conf. on Potential Energy Surfaces in Chemistry, ed. W.A. Lester Jr. (San Jose, IBM Research Laboratory) p.87.

5. Noble, C.J., 1982, Daresbury Laboratory Internal Report, DL/SCI/TM33T.

6. Baluja, K.L., Burke, P.G. and Morgan, L.A., 1982, Comput. Phys. Commun. 27, 299.

7. Gailitis, M., 1976, J. Phys. B, 9, 843.

8. Nesbet, R.K. and Noble, C.J., 1983, to be published.

9. Burke, P.G. and Salvini, S., 1983, to be published.

10. Kennerly, R.E., 1980, Phys. Rev. A, 21, 1876.

11. Burke, P.G., Noble, C.J. and Salvini, S., 1983, J. Phys. B, 16, L113.

12. Burke, P.G., Lamanna, U., Noble, C.J. and Salvini, S., 1983, to be published.

13. McLean, A.D. and Yoshimine, M., 1968, IBM J. Res. Dev. 12, 206.

14. Nesbet, R.K., 1969, Phys. Rev. 179, 60.

15. Collins, L.A., Robb, W.D. and Morrison, M.A., 1980, Phys. Rev. A, 21, 488.

16. Levin, D.A., Fliflet, A.W. and McKoy, V., 1980, Phys. Rev. A, 21, 1202.

17. Collins, L.A. and Schneider, B.I., 1983, Phys. Rev. A, 27, 101.

18. Schneider, B.I, and Collins, L.A., 1983, Phys. Rev. A, 28, 166.

19. Tennyson, J., Noble, C.J. and Salvini, S., 1983, to be published.

POLARIZATION AND CORRELATION EFFECTS IN LOW ENERGY ELECTRON MOLECULE COLLISIONS

B. I. Schneider and L. A. Collins
Theoretical Division
Los Alamos National Laboratory
Los Alamos, NM 87545

I. INTRODUCTION

The calculation of cross sections for the collision of electrons with molecular targets presents a considerably more difficult challenge than the corresponding problem for atomic systems. First, the non-spherically symmetric character of the electron-molecule interaction potential makes many of the numerical methods used for electron-atom scattering extremely difficult to apply and second, the nuclear motion introduces additional (ro-vibrational) degrees of freedom which can increase the number of channels far beyond the capabilities of even the most sophisticated computer.

Recent developments, which include the Linear Algebraic Method[1] (LAM), the Schwinger Variational Method[2] (SVM) and the R-matrix Method[3] (RMM), have relied on both discrete basis set (L^2) techniques, as well as more standard numerical approaches for their success. In the LAM, exchange and polarization are introduced into a set of coupled integral equations using basis set expansion and configuration interaction (CI) methods via separable expansions of the Feshbach optical potential. The resultant equations are reduced to a set of algebraic equations by the introduction of a set of Gauss quadratures.

In the next section we present a brief exposition of our optical potential technique and its implimentation within the LAM. Finally, we describe a number of recent and quite successful applications of the methodology to low energy electron diatomic collisions.

II. THEORY

Let us begin the discussion by decomposing the total interaction potential in channel c into three parts,

$$V_{cc'}(r,r',E) = V_{cc'}(r) \, \delta_{cc'} + K_{cc'}(r,r')\delta_{cc'} + U^{OPT}_{cc'}(r,r',E) \tag{1}$$

where

$$V_{cc'}(r)\delta_{cc'} \quad = \text{local, diagonal static interaction in channel c} \tag{2a}$$

$$K_{cc'}(r,r')\delta_{cc'} = \text{diagonal exchange interaction in channel c} \tag{2b}$$

$$U^{OPT}_{cc'}(r,r',E) \quad = \text{non-local, energy dependent optical potential} \tag{2c}$$

The local, diagonal, static interaction is easily treated within the LAM and need not be of much concern here. The traditional bottleneck in most treatments of electron scattering from molecules has been the exchange and polarization (correlation) potential. The basic reason for this is the non-local and non-separable form of equations (2b,c). Since exchange is a short range operator it was simultaneously suggested by us[4] and Rescigno and Orel[5] that separable expansions using a set of L^2 functions would be a very efficient approach to the construction of the exchange kernel. The method has the distinct advantage that standard integral and SCF techniques can be used to construct the required matrix elements. In addition the solution of the scattering equations reduces to the solution of a set of inhomogeneous equations, one for each L^2 function used in the expansion. In contrast to earlier L^2 studies which tried to use separable expansions for the entire interaction, convergence was quite rapid in function space. In addition the separable form improved execution times of the scattering codes by factors of 3 to 10. Encouraged by these findings we went on to explore the possibility of expanding the optical potential,

$$U^{OPT} = H_{PQ} (E-H_{QQ})^{-1} H_{QP} \tag{3}$$

as a sum of separable terms.[6] The projection operators P(Q) are defined as projectors onto the open (closed) channel portions of Hilbert space. In order to utilize the discrete basis methods developed for the exchange interaction, the continuum functions are replaced by a set of spin eigenfunctions in which all of the orbitals are pure L^2 Gaussian molecular orbitals. This allows us to employ standard bound state configuration interaction programs to construct the Hamiltonian matrix. The resulting matrix may be manipulated into the form of equation (3) and diagonalized to get

$$U^{OPT}(\vec{r},\vec{r}\,',E) = \sum_{\lambda} \psi_{\lambda}^{*}(\vec{r},E) \, \lambda(E) \, \psi_{\lambda}(\vec{r}\,',E) \tag{4}$$

which, aside from the energy dependence, is identical to that used to treat exchange. Of course, it is essential to choose proper one electron functions to account for polarization and to span the scattering continuum. We have employed the coupled Hartree-Fock method to get pseudostates which accurately reproduce the static polarizability of the target molecule. The continuum functions are mocked using a set of diffuse basis functions centered at the center of mass of the molecule. The method has been applied to low energy $e+H_2$ and N_2 collisions,[1] to the study of Feshbach resonances in $e+H_2^{+}$[7] scattering and most recently to very low energy $e+Li_2$ collisions. In general the results have been quite good and a short summary appears in the next section.

III. <u>RESULTS</u>

A. <u>$e+H_2$</u>

The elastic scattering of low energy (≤ 10 ev) electrons has been studied by a wide variety of experimental and theoretical techniques. Calculations within the static-exchange model are qualitatively incorrect in that the cross section shows no maximum and increases rather than decreases as the energy approaches zero. The inclusion of polarization either empirically or by <u>ab-initio</u> techniques greatly improves upon these results. Recent calculations using many-body optical potentials correct to second order have shown the power and generality of <u>ab-initio</u> methods. Other approaches, such as the adiabatic polarization technique, have also been successfully applied to $e+H_2$ collisions. Our method is similar to the many-body theory in that we construct an approximate optical potential. However we do not use perturbation theory but configuration interaction to construct the potential. The size of the CI matrix varied from about 350-450 configurations depending on the overall symmetry of the scattering state. In figure 1 we present a comparison of our calculated cross section with the experiments of Golden et. al.[8] and Dalba et. al.[9] Our calculations are in good agreement with experiment as well as the theoretical results of Morrison and Gibson.[10]

B.　e+N$_2$

The calculation of accurate, low energy e+N$_2$ cross sections has been hampered by the many electron nature of the target and the stronger interaction of the incident electron with the molecular charge cloud of the N$_2$. It is essential to be able to incorporate the correct polarizability of target in the scattering formalism to get quantitative results. This is difficult using perturbation theory correct to second order and hence seemed to be an interesting test of our new formalism. Just as in the case of e+H$_2$, we generated a set of pseudostate orbitals which reproduced the experimental static polarizability of the N$_2$ molecule to better than 10%. An additional set of diffuse orbitals were added to mock the p-space continuum in the optical potential. The resulting optical potential was constructed from about 850 configurations in which a single excitation of the N$_2$ molecular core was coupled to that of that of a scattered electron. The results for $^2\Sigma_g^+$ scattering are shown in table 1. The dramatic reduction of the cross section from the static-exchange result toward experiment is similar to that observed in e+H$_2$ collisions. Our calculations are in reasonable agreement with the experiments of Kennerly[11] as well as some recent calculations of Norcross and Padial[12] using a model potential which incorporated polarization and correlation.

C.　e+H$_2^+$

The scattering of electrons from ionic targets presents some problems which are different from those of neutral systems. In particular, excited ionic levels can trap electrons giving rise to a series of Rydberg or Feshbach resonances. The character of these resonances often changes from valence like to coulomb like depending on the character of the excited and trapped electron. We have performed a series of calculations on the $^1\Sigma_g^+$ resonances of H$_2$ using our optical potential model. These calculations were undertaken to help understand differences between the calculations of other workers using more approximate techniques. The results of our calculations and a comparison with Stieltjes, Kohn variational and close coupling calculations appears in fig. 2 and table II. Our calculations clearly demonstrate the need to include both valence and diffuse orbitals to describe this resonance series. The lowest member of the series is quite broad and very sensitive to the inclusion of correlation effects. The higher members are essentially uncorrelated and sensitive to the diffuse portion of the one-electron basis set. Thus it is essential to have a basis flexible enough to span a large portion of Hilbert space to describe the entire series correctly. None of the other calculations has this flexibility and were poor in either the low or high energy portion of the spectrum.

D.　e+Li$_2$

The calculation of low energy elastic Li$_2$ cross sections is an interesting but perplexing problem. Experiments by Bederson and Miller[13] over the energy range .5 to 10 ev show a very large cross section near zero which falls off reasonably rapidly with increasing energy. The huge magnitude of the cross section, the largest known for any homonuclear molecule, is similar in shape to that of the Li atom. Since the experimentalists were unable to carry out the experiments at lower energies the role of theoretical calculations would be of great help in elucidating the physics of the scattering process. The calculations we have undertaken were designed to give a good picture of the low energy $^2\Sigma_g^+$ and $^2\Sigma_u^+$ states of the negative ion. These calcu lations show that the $^2\Sigma_u^+$ negative ion state is bound with respect to the neutral from about R=4.5 a_o to larger internuclear distances. Since the equilibrium position of the neutral is R=5.05 a_o it is clear that the scattering cross section will be quite sensitive to vibrational effects. As the internuclear distance moves from larger to smaller R the $^2\Sigma_u^+$ state bound state pole moves into the complex plane and becomes a resonance. The situation with respect to the $^2\Sigma_g^+$ state is more complex. At internuclear distances around equilibrium of the neutral Li_2, the $^2\Sigma_g^+$ state of the ion is unbound. However, our scattering calculations show a marked change when we add the effects of polarization via an optical potential to the $^2\Sigma_g^+$ channel. In addition the effect seems to be larger at smaller internuclear distances. These effects, which are illustrated in figures 3-4, are not due to a resonance. The eigenphase sum in $^2\Sigma_g^+$ symmetry decreases as the energy increases from zero to a few tenths of an ev. The cross section on the other hand shows an enormous increase as the energy approaches zero. This suggests the presence of a virtual state in this symmetry not unlike that observed in other molecules with very attractive long-range potentials. The only disconcerting feature of the theoretical calculations when compared with experiment is the much more rapid fall off of the cross section with increasing energy. At present we do not know whether this is a defect in the calculation or a consequence of the uncertainty of the energy of the electron in the scattering experiment.

TABLE I. A comparison of low energy, static exchange (SE), effective optical potential (POL) and experimental (KENNERLY) $e+N_2$ cross sections.

k^2 (Ry)	δ (S.E.)	δ(Pol)	σ (S.E.)	σ(Pol)	σ(Exp)
.02	$-$ .3392	$-$.2330	72.28	33.45	31.10
.1	$-$ .7429	$-$.5749	60.57	40.50	38.09
.2	$-$ .9984	$-$.7691	48.72	36.73	
.3	-1.1732	$-$.9036	41.95	35.24	

TABLE II. Widths (Γ) and positions (E_r) for the lowest two $^1\Sigma_g$ resonances of e-H_2^+ scattering.

Method	E_r (eV)	Γ (eV)
Resonance 1		
Optical potential 1[a]	5.58	1.33
Optical potential 2	5.37	1.40
4 CC[b]	5.87	1.55
2 CC	6.04	1.38
Projection operator A [c]	5.57	1.32
Projection operator B		1.25
Kohn[d]	5.47	1.60
Projection operator[e]	5.53	1.35
Resonance 2		
Optical potential 2	9.92	0.151
4 CC	9.95	0.160
2 CC	9.96	0.148
Kohn	9.96	

[a]Present calculation effective optical potential bases 1 and 2.
[b]Four- and two-state close coupling
[c]Projection operator
[d]Kohn variational
[e]Projection operator

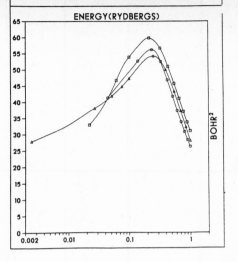

Figure 1

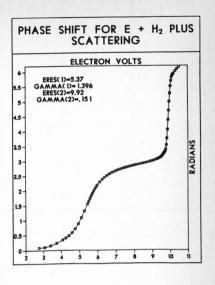

Figure 2

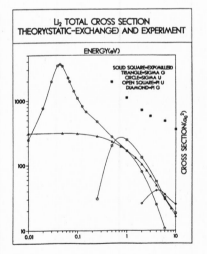

Figure 3

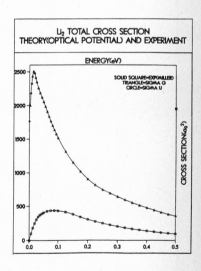

Figure 4

1. L. A. Collins and B. I. Schneider, Phys. Rev. A27, 101(1983).
2. R. R. Lucchese, D. K. Watson and V. McKoy, Phys. Rev. A22, 421(1980).
3. C. T. Noble, P. G. Burke and S. Salvini, J. Phys. B15, 3779(1982).
4. B. I. Schneider and L. A. Collins, Phys. Rev. A24, 1264(1981).
5. T. N. Rescigno and A. E. Orel, Phys. Rev. A23, 1134(1981).
6. B. I. Schneider and L. A. Collins, J. Phys. B 14, L101(1981).
7. B. I. Schneider and L. A. Collins, Phys. Rev. A28, 166(1983).
8. D. E. Golden, H. W. Bandel and J. A. Salerno, Phys. Rev. 146, 40(1966).
9. G. Dalba, P. Fornasini, I. Lazzizzera, G. Ranieri and A. Zecca, J. Phys. B, 2839(1980).
10. T. L. Gibson and M. A. Morrison, J. Phys. B, 14, 727(1981).
11. R. E. Kennerly, Phys. Rev. A21, 1876(1980).
12. D. W. Norcross and N. T. Padial, to be published.
13. T. M. Miller, A. Kasdan, B. Bederson, Phys. Rev. A25, 1777(1982).

THE SCATTERING OF SLOW ELECTRONS BY

POLYATOMIC MOLECULES

D. G. Thompson and A. K. Jain

Department of Applied Mathematics and Theoretical Physics
Queen's University, Belfast BT7 1NN, N. Ireland.

There has been considerable experimental activity in the area of electron scattering by polyatomic molecules (cf. for example this volume, and Ref. 1). This has not been matched either in quality or quantity by the theoretical work.

Some of our previous work[2] used interaction potentials with free-parameters chosen by appeal to experiment, but recently we have used a more ab-initio approach to study the following processes:

CH_4, NH_3: elastic and rotational excitation.[3,4,5]

H_2O, H_2S: elastic and rotational excitation, vibrational excitation of the symmetric stretch and bending modes[3,5,6,7]

We now think the calculations are sufficiently reliable to allow us to make a worthwhile contribution to the discussion of the physics of the scattering processes of these simple polyatomic systems.

Approximations

We first calculate a scattering amplitude $f(\hat{k},\hat{r};Q)$ for the scattering of an electron by a rigid fixed molecule. The electron is incident along $\underline{k}$ and scattered along $\hat{r}$, and Q represents the nuclear geometry. We obtain amplitudes at various Q and use the adiabatic approximation[8] to obtain an amplitude $f_{nn'}(\hat{k}.\hat{r})$ for excitation from nuclear state n to state n'.

$$f_{nn'}(\hat{k}.\hat{r}) = \int Z_{n'}^{*}(Q)\; f(\hat{k}.\hat{r};\, Q)\; Z_{n}(Q)\; dQ$$

$Z_n(Q)$ is the nuclear wavefunction appropriate for state n and in this work is either a rotation function or the product of rotation and vibration functions. This approximation is known to work well for rotational excitation except close to rotational thresholds. As we only consider excitation to low lying rotational states the approximation is quite satisfactory for this work. For vibration, as well as failing close to vibrational thresholds, the method also breaks down for processes proceeding through long lived resonances. Since the resonances reported here are all short lived the approximation is again quite satisfactory (For a general discussion of these matters see Herzenberg in Ref. 1).

As there are three vibrational degrees of freedom for H_2O and H_2S the determination of $f_{nn'}$ requires considerable computation. We have made the simplification of considering changes in $f(\hat{k}.\hat{r};\, Q)$ with respect to only one normal coordinate, variations in the other two being kept zero.

Scattering Model

We solved the following inhomogeneous differential equation for the electron scattering function $F(\underline{r})$.

$$(\nabla^2 + k^2 + V_S + V_E + V_P)F(\underline{r}) = \sum_i \lambda_i\; \phi_i$$

V_S is the static potential which can be calculated accurately from near-Hartree-Fock

single centre bound functions. The exchange interaction is introduced through the
local approximation of Hara[9],[10] (V_E), with also the possibility of orthogonalising
the continuum function F to the bound function ϕ_i of the same symmetry using the
Lagrange multiplier λ_i. There are no static-exchange calculations for polyatomics
with which we can test this approach, but previous work in atomic and diatomic systems
has proved to be very satisfactory[10]. V_p is a polarisation potential obtained using
the criterion of Temkin[11] and the approach of Pople and Schofield[12]. In electron-
atom scattering[13] this has proved a very good prescription for introducing non-adiabatic effects without arbitrary parameters chosen by appeal to experiment.

The approach described above - scattering by a fixed rigid molecule, plus adiabatic
approximation - is known to present special difficulties for polar molecules; the
differential cross section diverges as the scattering angle tends to zero and the
integrated cross section does not exist. There has been much discussion recently
on how to overcome this problem. A full radial frame transformation[14] is attractive
but extremely difficult to carry out. The angular frame transformation theory is
much simpler to apply and we have used the version discussed by Norcross and collaborators[15],[16]. We have calculated the differential cross section from the expression

$$I(\theta) = I^B + \sum_{\ell=0}^{L} (A_\ell - B_\ell) P_\ell(\cos\theta)$$

The $\Sigma A_\ell P_\ell$ is the usual expression obtained from a fixed nuclei scattering amplitude
plus transformation to laboratory frame of reference. The B_ℓ coefficients are
similar to the A_ℓ but calculated in the Born approximation using the long range
part of the potential. I^B is the Born approximation to the differential cross section. In this way we correct for the omission of high ℓ in the fixed nuclei approach
and because I^B is calculated in the laboratory frame I does not diverge as the
scattering angle tends to zero and the integral cross section does exist.

Results

1. CH_4, rotationally summed cross sections

(a) Integrated cross section (Fig. 1)

The cross section shows the well known Ramsauer-Townsend effect. It is a good test
of a theory to obtain agreement with experiment near the positions of the minimum
and maximum, since they are due to different scattering states, A_1 and T_2 respectively. The present results come out of this test remarkably well.

The T_2 maximum is mainly due to the incident electron partial wave $\ell = 2$, which
is the same as for the rare gases, argon, krypton and xenon which also exhibit the
Ramsauer-Townsend effect.

(b) Differential cross section

Another good test of theory is the differential cross section. We see, in Fig. 2
that there is good agreement between theory and experiment over a wide energy range;
even the oscillatory behaviour at the lower energies is reproduced by the theoretical
results.

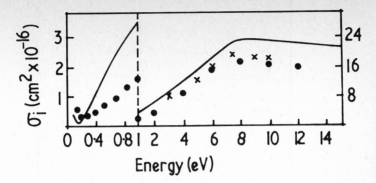

Fig. 1   CH$_4$ rotationally summed integrated cross section. ——, theory[3]; •, experiment, Barbarito et al[17] (agrees closely with Sohn et al[18] at low energies); x, experiment, Tanaka et al[19].

Fig 2 CH$_4$ rotationally summed differential cross section. ——, theory[3]; •, experiment, Tanaka et al[19]. (There are also experimental results at 5 ev by Rohr[20], agreeing closely with Tanaka et al).

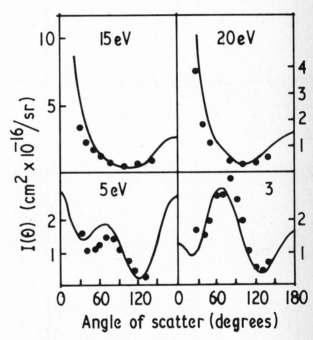

At 10 ev we can compare with the theoretical calculation of Abusalbi et al[21]. Their approach is to calculate the scattering by a rigid rotator which for CH$_4$ is a spherical top;   this involved solving the rotational close coupling equations in the laboratory frame.   Their model is similar in principal to that described above, but considerably different in detail;   it has a static potential obtained from a multicentre bound function, a local exchange term determined in the semi-classical approximation, and a local, energy dependent, non-adiabatic polarisation which is approximate but parameter free.

Agreement between the two calculations and experiment, is very good, as can be seen in Fig. 3.

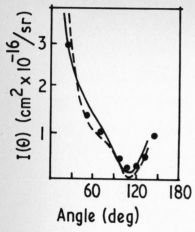

Fig. 3 CH_4 rotationally summed differential cross section,——, theory[5]; - - -, theory, Abusalbi et al[21]; •, experiment, Tanaka et al[19]. All at 10 ev.

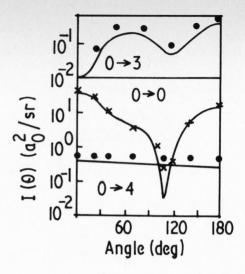

Fig. 4 CH_4 rotational excitation cross section at 10 ev.—— theory[5]; • and x, Abusalbi et al[21].

2. CH_4 rotational excitation cross sections at 10 ev (Fig 4)

Agreement between the two calculations for the $0 \rightarrow 3$ and $0 \rightarrow 4$ transitions is no longer good, differing by a factor of two. Rotational excitation cross sections are probably much more sensitive to the details of the polarisation potential.

3. NH_3 momentum transfer cross section (Fig. 5)

——, theory[4]
- - -, Born approx[4]
•, experiment,
Pack et al[22] for range
0.01-0.1 ev, compiled data
of Itikawa[23] for greater
energies.

Agreement between theory
and experiment is quite
good.

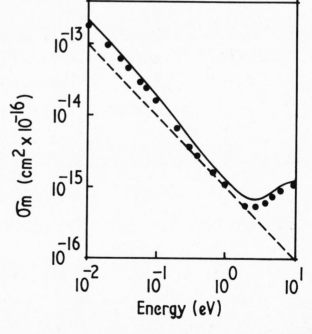

4. H_2O rotationally summed cross sections

(a) Integrated and momentum transfer cross sections (Fig 6)

————, theory[3]
•, experiment, Linder
cf Giantyrco and
Thompson[2]
x, experiment, Pack
et al[22]

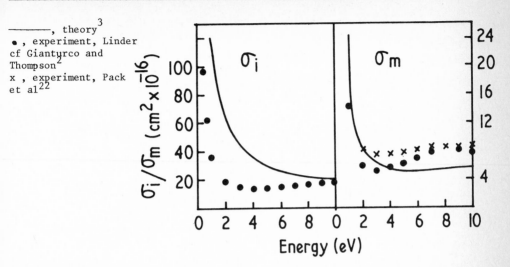

The disagreement between theory and experiment for the integrated cross section is probably due to the difficulty of describing correctly the small angle scattering; agreement is better for the momentum transfer cross section which is less sensitive to the small angle regime.

(b) Differential cross section (Fig 7)

————, theory, • experiment
as in Fig 6.

The agreement between theory and experiment is good at $90°$, but becomes increasingly poor, as expected, as the angle decreases.

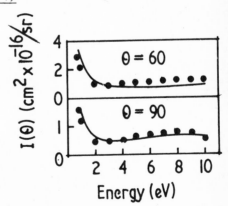

5. H_2O Rotational excitation

H_2O is an asymmetric top with energy levels $E_{J\tau}$ ($\tau = -J,\ldots+J$). Calculations have been carried out for several transitions including $(00 \rightarrow 00)$ and $(00 \rightarrow 10)$ while Jung et al[24] have measured cross sections for transitions ($\Delta J = 0$, $\Delta\tau = 0$) and ($\Delta J = +1$, $\Delta\tau = 0$) averaged over an initial rotational distribution (the gas temperature was sufficiently high for levels up to $J = 7$ to be significantly populated). Further calculations need to be done before a proper comparison between theory and experiment can be made.

6. H_2O vibrational excitation (Fig 8)

———, theory 4
● , experiment, Seng
and Linder[25].

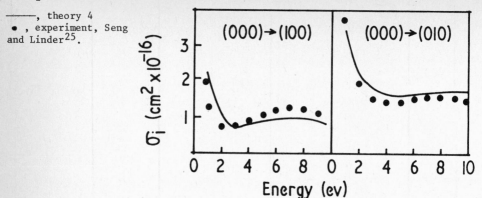

Agreement between theory and experiment is very good for the bending mode (010).
Note that for the symmetric stretch (100) part of the figure the experimental results
are for the sum of the symmetric and asymmetric (001) cross sections divided by four.

We have investigated the contribution of each scattering state to these cross sections.
We find that the low energy rise (leading to the threshold peak) is mainly due to the
A_1 state while at ∿7 ev it is the B_2 state (though with a large A_1 contribution) which
gives the major contribution. From dissociative attachment[26] experiments at this
energy there is clear evidence for a B_1 resonance state; however this is of the
Feshbach type associated with an excited state of H_2O and cannot be reproduced by the
model described above. Our conclusion about the importance of the B_2 state is repeated for H_2S.

7. H_2S rotationally summed cross sections

(a) Differential cross sections (Fig 9)

———, theory[7] ;[27,28]
●,experiment, Rohr
(all data divided by 2)

Agreement between theory
and the shape of the exp-
erimental data is very
good.

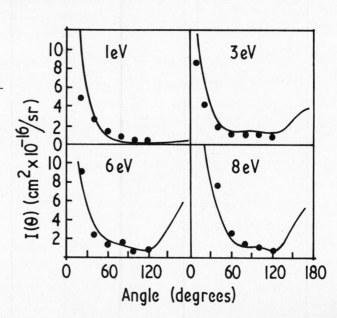

(b) Integral cross sections (Fig 10)

 theory[7]

(Note that the differential cross sections at $\theta = 90^\circ$ has the same shape as the integral cross section). The sharp rise at low energies is found to be predominantly due to A_1 scattering, the feature ~ 2 ev is due to B_2 and the feature at ~ 7 ev is due to A_1. From eigenphase sums it is possible to extract the following B_2 resonance parameters: $E = 2.2$ ev, $\Gamma = 1.3$ ev.

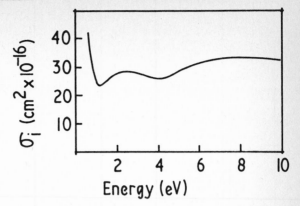

8. H_2S vibrational excitation – differential cross section

Note again the good agreement between theory and experimental shape as illustrated in Fig 11. The experimental data was originally interpreted in terms of an A_1 resonance state because of the isotropic behaviour between 20° and 120°. The present results are predominantly B_2. The effect of the B_2 resonance is seen clearly in Fig 12.

Fig 11

Fig 12 (100) only

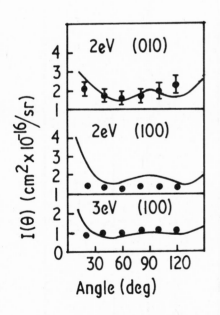

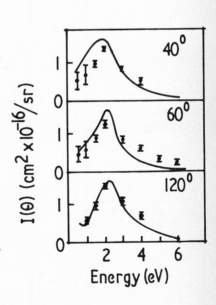

For both figures: ——, theory[7]; •, experiment, Rohr[27,28]. Note that in the (100) cases the experiment is for $\{(100) + (001)\}$ divided by 3.

Dissociative attachment experiments[29,30] at 2 ev have been interpreted in terms of an A_1 resonance state, at 5 ev in terms of a B_1 (Feshbach) resonance and at 8 ev in terms of an A_1 resonance. Obviously there is room for more work and discussion on these low energy scattering processes.

Conclusion

The scattering model presented here has obvious limitations but is a considerable advance on previous work in that good agreement is being obtained with experiment without recourse to the use of arbitrary parameters. We are now in a much better position to discuss the physics of the electron scattering processes of these simple polyatomic systems.

Acknowledgement

We are grateful to Dr. K. Rohr for sending us his experimental data on H_2S and also to Professor F. A. Gianturco for supplying us with the single centre code used to generate bound molecular orbitals.

References

1. Symposium on Electron-Molecule Collisions - Invited Papers, Univ. of Toyko, Sept. 1979;   eds I. Shimamura and M. Matsuzawa.
2. F. A. Gianturco and D. G. Thompson, 1980, J. Phys.B:At. Mol. Phys. 13 613-25.
3. Ashok Jain and D. G. Thompson, 1982, J. Phys.B:At. Mol. Phys. 15 L631-7.
4. Ashok Jain and D. G. Thompson, 1983, J. Phys.B:At. Mol. Phys. 16 2593-2607.
5. Ashok Jain and D. G. Thompson, 1983, J. Phys.B:At. Mol. Phys. 16 3077-3098.
6. Ashok Jain and D. G. Thompson, 1983, J. Phys.B:At. Mol. Phys. 16 L347-54.
7. Ashok Jain and D. G. Thompson, 1983, J. Phys.B:At. Mol. Phys. (In press).
8. D. M. Chase, 1956, Phys. Rev. 104 838.
9. S. Hara, 1967, J. Phys. Soc. Japan 22 710.
10. S. A. Salvini and D. G. Thompson 1981, J. Phys. B:At. Mol. Phys. 14 3797-3803.
11. A. Temkin, 1957, Phys. Rev. 107 1004.
12. J. A. Pople and P. Schofield 1957, Phil. Mag. 2 591-8.
13. D. G. Thompson, 1971, J. Phys. B:At. Mol. Phys. 4 468-82.
14. E. S. Chang and U. Fano, 1972, Phys. Rev. A6 173.
15. L. A. Collins and D. W. Norcross, 1978, Phys. Rev. A18 478.
16. N. T. Padial, D. W. Norcross and L. A. Collins, 1981, J. Phys. B:At. Mol. Phys. 14 2901-9.
17. E. Barbarito, M. Basta and M. Calicchio, 1979, J. Chem. Phys. 71 54-9.
18. W. Sohn, K. Jung and H. Ehrhardt, 1983, J. Phys. B:At. Mol. Phys. 16 891-901.
19. H. Tanaka, T. Okada, L. Boesten, T. Suzuki, T. Yamamoto and M. Kubo, 1982, J. Phys. B:At. Mol. Phys. 15 3305-20.
20. K. Rohr, 1980, J. Phys. B:At. Mol. Phys. 13 4897.
21. N. Abusalbi, R. A. Eades, Nam Tonny, D. Thirumalai, D. A. Dixon, D. G. Truhlar and M. Dupuis, 1983, J. Chem. Phys. 78 1213-27.
22. J. L. Pack, R. E. Voshall and A. V. Phelps, 1962, Phys. Rev. 127 2084.
23. Y. Itikawa, 1974, At. Data Nucl. Data Tables 14 1.
24. K. Jünk, Th. Antoni, R. Müller, K. H. Kochem and H. Ehrhardt, 1982, J. Phys. B:At. Mol. Phys. 15 3535-55.
25. G. Seng and F. Linder, 1976, J. Phys. B:At. Mol. Phys. 9 2539-51.
26. S. Trajmar and R. I. Hall, 1974, J. Phys. B:At. Mol. Phys. 7 L458-61.
27. K. Rohr, 1983, Private communication.
28. K. Rohr, 1978, J. Phys. B:At. Mol. Phys. 11 4109-17.
29. F. Fiquet-Fayard, J. P. Ziesel, R.Azria, M. Tronc and J. Chiari, 1972, J. Chem. Phys. 56 2540-8.
30. R. Azria, Y. Le Coat, G. Lefebre and D. Simon, 1979, J. Phys. B:At. Mol. Phys. 12 679-87.

CHARACTERISATION OF ELECTRONUCLEAR CORRELATIONS
IN LOW-ENERGY ELECTRON-MOLECULE DYNAMICS

M. Le Dourneuf and J.M. Launay
Observatoire de Paris, 92190 Meudon, France

I. INTRODUCTION

The description of electron-molecule dynamics faces two specific difficulties (1) the strong anisotropy and long range of the electrostatic interaction, due to the presence of two or more decentered nuclear singularities, and (2) the dynamical coupling to the nuclear degrees of freedom which leads to a large number of additional rovibrational excitation channels as well as rearrangment processes $(e+AB \rightarrow A+B^-)$.

The analysis of dominant physical mechanisms and their compact description by adapted representations constitutes a first goal for the theoretician aiming at the accurate prediction or interpretation of molecular dynamics.

A first step in the direction has been the definition of physically distinct regions of the electronuclear configuration space[1]. The huge mass difference between electrons and nuclei implies that the scattered electron gets accelerated while it penetrates into the molecular field, and tends rapidly to follow the nuclear motions. It leads to a Born-Oppenheimer factorization of the short range part of continuum molecular wavefunction, similar to the usual one for compact bound states. A practical implementation of the corresponding changes of representations refered to as the Frame Transformation Theory (FT), has progressively emerged: the range of adiabatic coupling can be defined quantitatively [2]; the local-adiabatic factorization of the AB^- compound's wavefunction is conveniently described by the R-matrix formalism and the asymptotic decoupling by quantum defect procedures . Accurate calculations for the rovibrational perturbations of Rydberg series in molecular spectra [4b,5] and resonant vibrational excitation[3b,6] have established the power of the FT sheme. A further progress in establishing the reality of the short range adiabatic electronuclear coupling and illustrating its theoretical simplifications, will be presented in §III through the actual plot of the electronuclear wavefunctions.

A second step in the compact representation of molecular continuum states is the definition of a rapidly-convergent expansion of the fixed-nuclei electronic continuum orbital on an adiabatic angular basis, which diagonalizes the fixed R,r hamiltonian $V_{eff}(R,r;\hat{r})=V(R,r;\hat{r})+\vec{\ell}^2/2mr^2$ for each radial configuration[7]. The importance of centrifugal effects had been naturally exploited in atoms: the approximate spherical symmetry leads to an approximate decoupling of spherical modes, and their radial evolution is determined by the effective potentials $V_{eff}(r)$, the visualisation of which has provided insight into the systematics of atomic dynamics[8]. In molecules, the importance of centrifugal effects, which had been observed experimentally (pure-ℓ character of Rydberg series, and angular distributions[9]), is now parallely introduced by the theory. In §II, we show how the adiabatic angular expansion leads to an accurate description of the fixed-nuclei electronic scattering in terms of a few locally open channels: typically 1-2 for homonuclear systems, 2-5 for heteronuclear systems. This compact representation provides a systematic tool to predict shape resonances by direct inspection of single dimension effective adiabatic potentials[10]. A complementary element in the visualisation of trapping mechanisms of scattered electrons by the molecular field is of course provided by the plot of electronic wavefunctions.

II. ELECTRON-MOLECULE FIXED-NUCLEI SCATTERING

Owing to the heaviness of the target nuclei, the scattering by a fixed-nuclei target is a good approximation in most cases, or the first step of the short range part of a full electronuclear calculation in the case of a resonant trapping of the scattered electron. After a brief review of the formalism (§II A), we consider two typical examples of homonuclear $(e+N_2)$ and heteronuclear$(e+HCl)$ diatomic systems.

II A. Adiabatic angular close-coupling

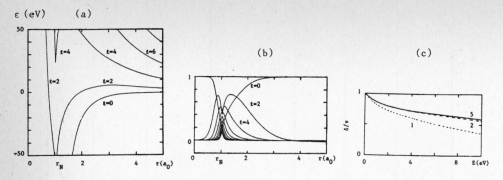

Fig. 1: $e(\sigma_g) + N_2$ scattering at R_{eq}
(a) Adiabatic potentials $\varepsilon_\ell^{\sigma_g}$
(b) ℓ-components of $\phi_0^{\sigma_g}$
(c) convergence of the eigenphase sum with the number of adiabatic angular modes included in the expansion (Eq.3.).

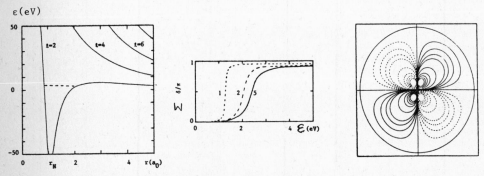

Fig. 2: $e(\pi_g) + N_2$ scattering at R_{eq}
(a) Adiabatic potentials $\varepsilon_\ell^{\pi_g}(R_{eq};r)$ with the dashed line on the $\ell=2$ curve indicating the 2.4 eV shape resonance.
(b) convergence of the eigenphase sum with the number of adiabatic angular modes.
(c) Fixed nuclei electronic wavefunction $\Phi^{\pi_g}(R_{eq};r,\theta,\phi=0)$ at 2.4 eV. Hached contours are negative.

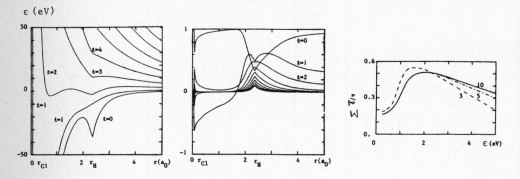

Fig. 3: $e(\sigma) + HC\ell$ scattering at R_{eq}
(a) Adiabatic potentials $\varepsilon_\ell^\sigma(R_{eq};r)$
(b) ℓ-components of ϕ_0^σ
(c) convergence of the eigenphase sum $\sum \bar{\delta}_\alpha$ in the dipole representation

Starting from a realistic local approximation of the interaction potential, discussed in detail in ref[10] of the present volume

$$V(R;\vec{r}) = V^{stat}(R,\vec{r}) + V^{pol}(R,\vec{r}) + V^{exc}(\vec{r}) \qquad (1)$$

we define the adiabatic angular basis by diagonalizing the fixed (R,r) hamiltonian within each irreducible representation of the symmetry group of the molecule ($\lambda = \sigma, \pi \ldots$ for a diatomic system)

$$\left[\frac{\vec{\ell}^2}{2mr^2} + V(R,r;\hat{r}) \right] \phi_\ell^\lambda (R,r;\hat{r}) = \varepsilon_\ell^\lambda (R,r) \, \phi_\ell^\lambda (R,r;\hat{r}) \qquad (2)$$

The adiabatic angular functions ϕ_ℓ^λ, which depend parametrically on the internuclear distance R and the radial distance r to the molecular center of mass, reduce to the usual spherical harmonics $Y_\ell^\lambda(\hat{r})$ whenever $\vec{\ell}^2/2mr^2 \gg V$, that is for $r \to 0, r \to \infty$ (non polar molecules) and at all r for large ℓ. The adiabatic index coïncides in these limits with the orbital angular momentum quantum number. However, the lower ϕ_ℓ^λ's get appreciably distorted near the nuclear singularities, and for polar molecules, the ϕ_ℓ^λ's converge at large r to the dipolar harmonics, i.e. the eigenfunctions of $\vec{\ell}^2 - 2\vec{D}.\hat{r}$.

The adiabatic expansion of the continuum orbital:

$$\phi^\lambda (R;\vec{r}) = \sum_\ell \phi_\ell^\lambda (R,r;\hat{r}) \; \frac{f_\ell^\lambda (R;r)}{r}$$

defines radial orbitals f_ℓ^λ, which depend parametrically on (R,r) through the adiabatic basis. In the fixed **nuclei** scheme, the f_ℓ^λ's are determined by the standard non adiabatic coupled equations:

$$\left[\sum_{\ell'} - \frac{1}{2m} \left(\frac{d}{dr} + P^\lambda \right)^2_{\ell\ell'} + \left(\varepsilon_\ell^\lambda (R;r) - \varepsilon \right) \delta_{\ell\ell'} \right] f_\ell^\lambda (R;r) = 0 \qquad (4a)$$

with the electronically non adiabatic couplings

$$P_{\ell\ell'}^\lambda (R;r) = \int \phi_\ell^\lambda (R,r;\hat{r}) \; \frac{\partial}{\partial r} \; \phi_{\ell'}^\lambda (R,r;\hat{r}) \; d\hat{r} \qquad (4b)$$

induced by the variation of the adiabatic angular basis. Therefore, the f_ℓ^λ's evolve along the adiabatic effective potentials $\varepsilon_\ell^\lambda (R;r)$, with transitions induced by the $P_{\ell\ell'}^\lambda$'s, the influence of which, being concentrated near the nuclear singularities, is reduced by the large centrifugal effects near the nuclei. These strong centrifugal effects, combined with the effective repulsion induced by the Pauli exclusion principle[10] (enforced by the explicit orthogonalisation of ψ^λ to the molecular closed shells of symmetry λ), imply that, in practice, converged results are obtained with a few adiabatic channels: only those which are locally open and whose energies depart appreciably from the corresponding centrifugal potentials.

Note that, if the calculation and visualisation of non adiabatic couplings help in estimating quickly their influence, their sharp variation close to the nuclear singularities is inconvenient in practical calculations, where the adiabatic angular expansion (continuous frame transformation) is replaced by a constant angular expansion adapted to radial sectors (discrete frame transformation)[4d,11].

II B. Example of an homonuclear diatomic system: e + N₂

Fig.1 illustrates the scattering of a penetrating σ_g symmetry. Only the 3 lowest adiabatic energies ($\ell = 0, 2, 4$) differ notably from the pure centrifugal potentials (fig.1a). The expansion of the ϕ_ℓ^λ on the spherical basis Y_ℓ^λ illustrates the sharp localisation of the ℓ mixing on the nuclear singularities (fig.1b). Converged eigenphase sums are obtained up to 10eV by including only the first two locally open adiabatic modes (fig.1c).

Fig.2 illustrates the alternative case of a less penetrating, but resonant, π_g scattering symmetry. Now, only the first $\ell = 2$ adiabatic potential penetrates deeply into the molecule, and its shape suggests immediately the trapping mechanism responsible for the well known $N_2^-(^2\Pi_g)$ shape resonance around 2.4 eV (fig.2a). A single channel scattering calculation does reproduce the resonance within 1eV of the converged result (fig.2b). Convergence to the exact resonance energy proceeds from below because of the kinetic character of non adiabatic effects, but it is much faster than in the usual spherical harmonics expansion (limit reached with 3 adiabatic orbitals, instead of 10-15 spherical harmonics). Fig.2c shows the

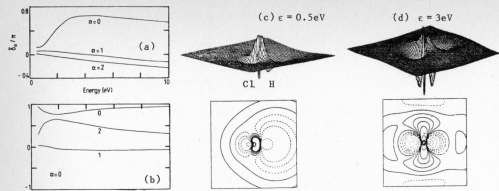

Fig.4: e(σ)+HCl scattering (a) Energy dependence of the eigenphases $\overline{\delta_\alpha}$ in the dipole field representation, with 3 adiabatic channels in the expansion (eq.3).
(b) Energy dependence of the expansion of the $\alpha=0$ eigenchannel on the first 3 adiabatic angular modes. (c) Wavefunction of the $\alpha=0$ eigenchannel at $\varepsilon=0.5$eV. Dotted contours have negative amplitude .(d) same as(c) at $\varepsilon=3$eV.

Fig.5: The four regions of configuration space for electron-molecule scattering. r is the electron-center of mass distance and R the internuclear separation.

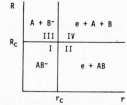

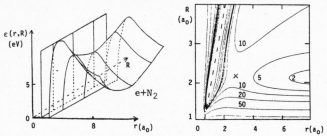

Fig.6: The lowest $^2\Pi_g$ electronuclear potential energy surface for N_2^-. Energy countours are in eV.

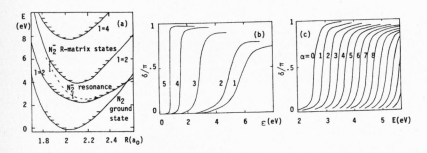

Fig.7: (a) $N_2^-(^2\Pi_g)$ fixed-nuclei electronic R-matrix states and resonance position.
(b) The fixed-nuclei phaseshift $\delta(\varepsilon;R)$, curves 1 to 5 correspond to R=1.744, 1.868, 2.068, 2.268, 2.391 a_o respectively.
(c) Electronuclear eigenphaseshifts $\delta_\alpha(E)$.

resonant wave function: its dominant $\ell = 2$ character is distorted only locally near the nuclei.

II C. Example of an heteronuclear diatomic system: e + HCl

Fig.3,4 illustrate the more complicated heteronuclear case. The adiabatic energies for the σ symmetry (fig.3a) reflect the presence of 2 separate wells at the radial distances associated with the 2 nuclear singularities.Even and odd ℓ's are coupled, though not very tightly, because of the predominance of the Cl field. Only the 3 lowest curves show wells and the $\ell = 1$ curve is particularly weakly influenced by the H singularity. The convergence of adiabatic modes to dipolar harmonics is apparent in Fig. 3b for $r \geqslant 5a_o$. Although non adiabatic effects are expected from the artificial jump of the lower adiabatic modes from one nuclear well to the other, it is important to note that the coupling of the first 3, locally open, adiabatic channels suffices to describe semi-quantitatively the scattering up to 10 eV, and 5 adiabatic channels give fully-converged results (Fig.3c).

The strong coupling between the first 3 channels (and mainly $\ell = 0,2$) is reflected by the strong energy dependence of the α=0 eigenphase and the channel composition of the resonant scattering eigenchannel [12] (α=0), defined w.r.t. a dipolar asymptotic basis (Fig.4a-4b). The strong angular variation of the corresponding wavefunctions between threshold (ε=0.5eV,Fig.4c.), and the resonant energy (ε=3eV,Fig.4d), provides a preliminary insight on the physical mechanism responsible for the enhancement of vibrational excitation observed experimentally near threshold and around 3eV [13]. Thus, the 0.5 eV wavefunction, which corresponds to the penetration of the first mode (α=0)on the H side, illustrates the virtual-state mechanism proposed by Herzenberg [14] to explain the threshold vibrational peak. At higher energy(fig4d), the strong mixing with the $\ell = 2$ channel which is reminiscent of the $\ell = 2$ shape resonance in the Ar isoelectronic species[7], leads to an axial localisation of the trapped electron near the two nuclear singularities and to an enhanced transfer of impulsion to the nuclear vibration.

This preliminary interpretation of electronuclear correlations should be completed by allowing and visualizing the correlated electronuclear motion, following a technology, which we will now present in the simpler homonuclear case.

III. RESONANT COUPLING TO THE NUCLEAR MOTION

After a short presentation of the doubly-adiabatic R-matrix formalism of coupled electrovibrational motion (§III A), we will present extensive analysis of the electronuclear correlations responsible for resonant vibrational excitation $\left[N_2^- \ (^2\Pi_g) \text{shape resonance} \right]$ in §III B and dissociative attachement $\left[H_2^- \ (^2\Sigma_u) \text{ first} \right.$ process around 4eV$\left. \right]$ in §III C.

III A. Doubly-adiabatic R-matrix description of short range electronuclear correlations

Relaxing the internuclear elongation R, but still neglecting the molecular rotation during the collision time, the electronuclear wavefunction $\Psi^\lambda(R,\vec{r})$ satisfies the following Schrödinger equation:

$$\left[-\frac{1}{2MR}\frac{\partial^2}{\partial R^2}R - \frac{1}{2mr}\frac{\partial^2}{\partial r^2}r + \frac{\vec{\ell}^2}{2mr^2} + V(R,\vec{r}) + V_{AB}(R) - E \right] \Psi^\lambda(R,\vec{r}) = 0 \tag{5}$$

where M is the reduced mass for the nuclear motion and $V_{AB}(R)$ is the electronic energy of the target ground state. Depending on the relative magnitudes of R and r, four regions of configuration space can be isolated (Fig.5),corresponding to different physical conditions.

While region IV (three body dissociation) is forbidden at low-energy, regions II and III are the electronic and nuclear fragmentation zones, well described by expansion on asymptotic decoupled bases, i.e.:

- for region II(e+AB): $\Psi^\lambda_{v\ell k_v}(R,\vec{r}) = \dfrac{\chi_v^{AB}(R)}{R} \ Y_\ell^\lambda(\hat{r}) \ \dfrac{f_{v\ell}^\lambda(k_v r)}{r}$ \hfill (6)

- for region III(A+B⁻) : $\Psi^\lambda_K(R,\vec{r}) = \Phi^{AB^-}(R;\vec{r}) \ f^{AB^-}(KR)/R$ \hfill (7)

where, if not obvious from the notations, $\chi_v^{AB}(R)$ are vibrational eigenstates of the AB molecule, $\Phi^{AB^-}(R,\vec{r})$ the Born-Oppenheimer electronic state in the nuclear fragmentation channel A+B⁻, $f_{v\ell}^\lambda(k_v r)$ and $f^{AB^-}(KR)$ decoupled radial solutions

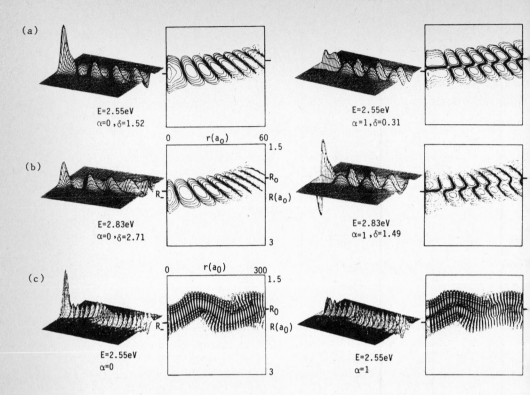

Fig. 8: The $N_2^-(^2\Pi_g)$ electronuclear eigenmodes $F(R,r)$
a. Eigenchannels $\alpha=0,1$ at the first vibrational resonance $E=2.5$ eV.
r lies between 0 and $60a_0$, and R between 1.5 and $3a_0$. δ is the phaseshift
in radians.
The equilibrium positions of N_2^- and N_2 are respectively indicated on the
left (R_-) and right (R_0) hand sides of the contour plots.
b. Eigenchannels $\alpha=0,1$ at the second vibrational resonance $E=2.8$ eV.
c. Eigenchannels $\alpha=0,1$ at $E=2.55$ eV, but r between 0 and $300a_0$.

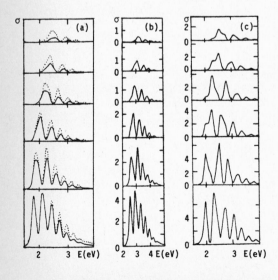

Fig. 9:
Comparison of present results with
experiment and with previous theoretical results for vibrational excitation cross sections $\sigma_{0\to v}$ (v=1 to 6
from bottom to top). Cross sections
are in $\AA^2$.
(a) Full line: boomerang model of Dubé
and Herzenberg (1979)[18]
Dotted line: experiment of
Ehrardt and Willmann (1967)[19]
(b) Present results.
(c) Multicenter R-matrix theory
of Schneider et al. (1979b)[6]

corresponding to moments k_ν and K in the asymptotic electronic and nuclear channels.

Region I ($r \ll r_c, R \ll R_c$) is the region of strong electronuclear correlation, where the Born-Oppenheimer factorisation of the electronuclear functions locally prevails. A discrete basis of doubly adiabatic electronuclear states:

$$\psi_{\ell n \nu}^\lambda (R, \vec{r}) = \Phi_\ell^\lambda (R, r; \hat{r}) \frac{f_n^{\lambda \ell}(R; r)}{r} \frac{\chi_\nu^{\lambda \ell n}(R)}{R} \tag{8}$$

is defined, using the adiabatic angular basis Φ_ℓ^λ (eq.1) and solving successively the fixed-nuclei electronic and nuclear R-matrix radial equations:

$$\{ \frac{1}{2m} \left[\frac{\partial^2}{\partial r^2} + L(b, r_c; r) \right] + \varepsilon_\ell^\lambda (R, r) \} f_n^{\lambda \ell} (R; r) = \varepsilon_{\ell n}^\lambda (R) \; f_n^{\lambda \ell} (R; r) \tag{9a}$$

$$\{ \frac{1}{2M} \left[-\frac{d^2}{dR^2} + L(B, R_c; R) \right] + \varepsilon_{\ell n}^\lambda (R) + V_{AB}(R) \} \chi_\nu^{\lambda \ell n}(R) = E_{\ell n \nu}^\lambda \; \chi_\nu^{\lambda \ell n} (R) \tag{9b}$$

where the L's are singular surface operators [15], which enforce a fixed logarithmic derivative at the boundary of the finite radial range

$$L (b, r_c; r) = \delta (r - r_c) \left[\frac{\partial}{\partial r} - b \right] \quad \text{,and similarly for } L(B, R_c; R) \tag{9c}$$

Non adiabatic effets, due to the R-variation of the electronic radial functions $f_n^{\lambda \ell}$ (R;r), are made completely negligible by the R-matrix discretization of short-range electronic eigenmodes and energies $\varepsilon_{\ell n}^\lambda$ (R). However, those due to the r/R variation of the electronic angular modes Φ_ℓ^λ (R,r;$\hat{r}$), can only be neglected in a semi-quantitative description of shape resonances in homonuclear systems §IIB, which we will do in the following to concentrate on the graphical analysis of the important radial correlations. These non-adiabatic effets in r/R must however be included in the calculation of the fixed-nuclei electronic R-matrix states for accurate results on homonuclear systems and even qualitative ones on heteronuclear systems. The uncoupled electronic eq.9a is then replaced by a few coupled equations according to the formalism developped in §II A.

Once the Born-Oppenheimer R-matrix basis is determined, the globally non adiabatic electronuclear scattering functions can be obtained by imposition of suitable asymptotic conditions and matching between the compound (I) and the fragmentations (II,III,IV) regions.

III B. Resonant vibrational excitation of N_2 by electron impact.

Fig.6 shows the lowest angularly-adiabatic electronuclear potential surface $\varepsilon_2^{\pi g} (r,R) + V_{N_2}(R)$ for the N_2^- ($^2\Pi_g$) system. It provides the first graphical display of the origin of the electronuclear correlations responsible for enhanced vibrational excitation (Fig.9). The incident $e + N_2$ arrangment tunnels through the centrifugal barrier and gets trapped into the narrow, but deep, ditch of nuclear singularities (r=R/2,bottom around-50eV). The ditch gets narrower at larger R,preventing the opening of the dissociative attachment channel, and the system bounces back with possible vibrational excitation.

Fig.7 shows intermediaries of the calculation.Fig.7a visualises the N_2 potential curve, the N_2^- R-matrix potentials (for null-derivative at $r_c = 8 a_0$), their vibrational levels (null-derivative at $R_c = 3 a_0$), and the N_2^- fixed-nuclei resonance position deduced from the analysis of the fixed-nuclei electronic phase shift (Fig.7b). The non-parallelism of the first 2 R-matrix states to the N_2 potential results from the strong correlation of the trapped electron with the nuclear motion. Fig.7c shows the electronuclear eigenphases, whose energy variation reproduces the vibrational structure of the N_2^- resonant state (Fig.7a).

An extensive characterization of the electronuclear correlations is given by the S-matrix eigenmodes, usually refered to as the eigenchannels [12], and characterized by the same phaseshift in all vibrational channels. Fig.8 shows the first two eigenmodes, which have a simple nodal characterisation at all energies. At short distances, the nodal structure is typical of an adiabatic factorisation of the compound wavefunction:oblique electronic node associated to the fixed-nuclei local wavenumber k(R;r), straight nuclear antinode or node, which coincides with $R = R_-$ at the resonance energies (E=2.55 eV for $\alpha = 0$, E=2.8 eV for $\alpha = 1$), where the amplitude of the wavefunction increases rapidly in the compound region. At resonance, the eigenmodes coincide then with the longest-lived eigenmode of the collisional li-

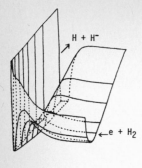
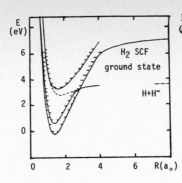
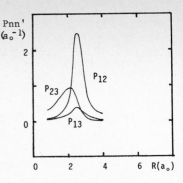

Fig.10:The lowest $^2\Sigma_u$ electronuclear potential energy surface for H_2^- .

Fig.11a:$H_2^-(^2\Sigma_u)$fixed-nuclei electronic R-matrix states and resonance position (dashed line).

Fig.11b:Non adiabatic couplings due to the R-variation of electronic R-matrix states.

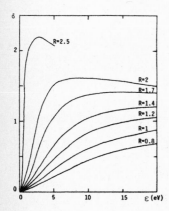
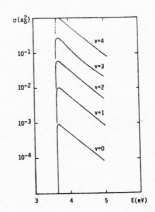

Fig.12:The fixed-nuclei phaseshift $\delta(\varepsilon;R)$ as a function of $R(a_o)$ for $H_2^-(^2\Sigma_u)$.

Fig.13: Dissociative attachment cross-section from $H_2(v)$.

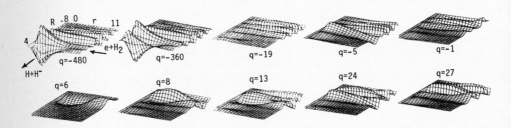

Fig.14: The $H_2^-(^2\Sigma_u)$ electronuclear eigenmodes of the lifetime matrix at 4eV. r lies between 0 and $11a_o$,R between .8 and $4a_o$. q is the lifetime value.Note the two dissociating modes with q = - 480 and q = -360.

fe-time matrix $Q = -i\hbar\, S^{-1} dS/dE$ [16]. But, in contrast to these, all eigenchannels keep a sharp characterisation at all energies.

This characterization results from the combination of the adiabatic character of the electronuclear correlations and the effective local character of the electronic detachment process [17]. We have shown in ref 11 that the α eigenchannels have then an approximately factorized form:

$$F\ (E = E_\alpha^\delta \ ;R,r) = f(\varepsilon_\delta(R),R;r)\ \chi_\alpha\ (U_\delta^-(R);R) \qquad (10)$$

where χ_α is the αth vibrational level of energy $E_\alpha^\delta = E$ in the electronic potential $U_\delta^-(R) = V_{N_2}(R) + \varepsilon_\delta(R)$ defined by the functional relationship between the fixed-nuclei electronic phase and energy (Fig 7b). This leads to an adiabatic partitioning of energy between the continuum electron and the nuclei. Assuming a Breit-Wigner form for the resonant phase-shift,

$$U_\delta^-(R) = U_{res}\ (R) - \frac{1}{2}\ \Gamma(R)\ \cotan\ \delta \qquad (11)$$

the monotonic shift of the short range vibrational nodes with E results from a shift of the minimum of U_δ^- (R) towards increasing R for increasing δ, due to the decrease of Γ with R.

Fig.8c shows the evolution of the eigenchannels up to large electronic distance ($r \leqslant 300 a_0$). The highly coherent electronuclear motion spreads out, due to the gradual detuning of the electronic phases $k_v r$. The oscillation of wavelength $V_e \cdot T_v$ (V_e=electronic velocity, T_v=vibrational period for N_2) $\leqslant 300 a_0$, is induced by the electronic transfer of impulsion within the correlation zone.

Finally, Fig.9 shows that the experimental oscillatory structure in vibrational excitation cross sections is reproduced semi-quantitatively by the present calculation, whose results are nearly as good as those of an extensive multicenter R-matrix calculation[3b,6].

III C. Dissociative attachment of H_2

Fig.10 shows the first adiabatic electronuclear surface for the H_2^- ($^2\Sigma_u$) system. Fig.11a shows the corresponding R-matrix fixed-nuclei electronic energies $\varepsilon_n(R)$ and electronuclear levels $E_{n\nu}$ for a complex region $r_c = 11 a_0$, $R_c = 4 a_0$ and a null derivative at both boundaries. The first electronic R-matrix state converges to the H_2^-($^2\Sigma_u$) electronically bound state after the stabilisation point $r_s = 3 a_0$. Its higher electronuclear states have non-zero amplitude at the nuclear boundary and describe the dissociative attachment process. Fig.11b illustrates the smallness of non-adiabatic effects associated with the radial electronic motion in the region of the complex (I on Fig. 5), establishing for the first time the relevance of the adiabatic R-matrix approach for dissociative attachment.

The fixed-nuclei electronic phaseshifts, displayed on Fig.12, show that the electronic trapping is very small in the Franck-Condon region, but the resonance gets narrower and becomes a bound state at large R. This explains the rapid increase of dissociative attachment cross-sections, with the increasing initial vibrational level of the target (Fig.13).

The properties of the complex are illustrated on Fig.14, which shows the collisional life-time eigenmodes in the region of the complex. There are 10 open channels (9 vibrational and 1 dissociative) at 4eV, but only 2 Q-eigenmodes form a resonant complex which can simultaneously detach and dissociate. The corresponding wavefunctions have a non-resonant behavior in the Franck-Condon region of H_2 ($R=1.4 a_0$), in contrast to the N_2^- case. The flux leading to dissociation climbs on the potential ridge which separates the electronic and nuclear valleys. The influence of initial vibrational excitation on the reaction probability is obvious from the structure of the dissociating eigenmodes on the electronic boundary. Note that the eigenchannels (S-eigenmodes) do not show such a sharp characterization, and are therefore less useful in displaying the mechanism of dissociative attachment.

This pilot calculation of a dissociative attachment process proves to be very instructive, in spite of its crudeness. Improvements of the present picture are under investigation, by relaxing the main approximations (Hartree- Fock description of H_2 for all R, statistical approximation of exchange, adiabatic factorization of the electronic angular part of the wavefunction).

CONCLUSION

In this paper, we have shown that the light mass of the electron and the strength of its interaction with a molecular target limit the number of eigenmodes which contribute effectively to low energy electron-molecule dynamics by favouring adiabatic factorizations of fast motions.The Born-Oppenheimer factorization of fixed-nuclei electronic and nuclear motions, familiar in bound molecular states, has been extended in two directions:the short range part of the electronic continuum (R-matrix and frame-transformation techniques) and the angular electronic motion, which adjusts to the molecular anisotropy at each distance of approach (adiabatic angular expansion).

These successive adiabatic factorizations lead to a sequence of simple one dimensional motions in effective potentials,allowing a visual prediction of resonances and a detailed estimation of coupling mechanisms.

These approximate factorizations facilitate also the visualisation of the system's wavefunction by low-dimensional plots .The properties of correlated electronuclear continuum states emerge from the analysis of suitable eigenmodes.In particular, we have shown that the strong and short-range electronuclear coupling in N_2^- ($^2\Pi_g$) around 2eV leads to a simple nodal structure of all the phase eigenmodes and a unique visualisation of resonant nuclear excitation by electron impact.In addition, the life-time eigenmodes permit to isolate resonant states,which are strongly localized in the complex zone and to visualize the mechanisms of rearrangement clarifying for example the importance of initial vibrational excitation in the $e+H_2 \rightarrow H+H^-$ dissociative attachment process around 4eV.

REFERENCES

1. E.S.Chang and U.Fano Phys.Review A6,173 (1972)
2. J.M.Launay and M.Le Dourneuf,Abstracts,XIIth ICPEAC,ed.S.Datz (Gatlingbürg,USA)P.933
3. a. B.L.Schneider,M.Le Dourneuf and P.G.Burke,J.Phys.B12 L365 (1979)
   b. M.Le Dourneuf,L.Vo Ky and B.I.Schneider,in'Electron-atom and Electron-molecule Collisions',ed.J.Hinze,Plenum Press 1983,P.135
4. a. C.H.Greene,U.Fano and G.Strinati,Phys.Rev.A19 1485 (1979)
   b. C.Jungen and O.Atabek,J.Chem.Phys.66 5584 (1977)
   c. L.Vo Ky and M.Le Dourneuf,in 'Electronic and Atomic Collisions',Invited Papers of the XIth ICPEAC 979,ed.by Oda and Takayanagi,North Holland 1980,P.751
   d. L.Vo Ky,M.Le Dourneuf and J.M.Launay,in'Electron-Atom and Electron-molecule Collisions',ed.J.Hinze,Plenum Press 1983 P.161
5. Ch.Jungen,in'Electronic and Atomic collisions', ed.S.Datz,North Holland 1982, P.455 A.Guisti-Suzor,id,P.381
6. B.I.Schneider,M.Le Dourneuf andL.Vo Ky,Phys.Rev.Lett.43 1927 (1979)
7. M.Le Dourneuf,L.Vo Ky and J.M.Launay J.Phys.B15 L 685 (1982)
8. U.Fano and J.W.Cooper,Rev.Mod.Phys.40 441 (1968)
9. L.Malegat,M.Le Dourneuf and M.Tronc,this volume
10. B.J.Austin,V.Heine and L.J.Shem,Phys.Rev.127276 (1962)
11. M.Le Dourneuf,J.M.Launay and L.Vo Ky,in 'Recent Developments in Electron-Atom and Electron-Molecule Collision Processes',ed.W.Eissner,Daresbury Laboratory Report DL/SCI/R 18 P.70
12. U.Fano and C.M.Lee Phys.Rev.Lett. 31,1573 (1973) D.Loomba,S.Wallace,D.Dill and J.L.Dehmer,J.Chem.Phys.75,4546 (1981)
13. K.Rohr and F.Linder ,J.Phys. B10 L200 (1975)
14. A.Herzenberg,in'Electronic and Atomic Collisions',ed.GWatel,North Holland 1978 ,P.1
15. C.Bloch,Nucl.Phys.4,503 (1957)
16. F.T.Smith,Phys.Rev.118,349 (1960)
17. J.N.Bardsley,A.Herzenberg and F.Mandl,Proc.Phys.Soc.89,321 (1966)
    J.N.Bardsley,J.Phys. B1,321 (1968)
18. L.Dubé and A.Herzenberg ,Phys.Rev.A20, 194 (1979)
19. H.Ehrhardt and K.Willmann,Z.Phys.204,462 (1967)

CALCULATION OF EFFECTIVE POTENTIALS FOR THE STUDY OF SHAPE RESONANCES IN OPEN SHELL DIATOMICS

L. Malegat[*], M. Le Dourneuf[**], M. Tronc[*].

[*]Laboratoire de Chimie Physique, Université ParisVI, 75005 Paris
[**] Observatoire de Paris, 92190 Meudon, France.

A. INTRODUCTION

Shape resonances play a dominant role in low energy electron-molecule dynamics. Among the numerous ones observed in photoionization, elastic or vibrationally ine- lastic electron scattering, a few cases are now well understood. It has been confir- med theoretically that they originate in a particular shape of the electron-target interaction potential as suspected from experiment. However, we are still far from understanding the systematics of experimental data, even in the simple case of the first row diatomics like N_2 , CO, NO, O_2.[1] At this stage, there is still room for a theoretical tool that would satisfy the following requirements:
1°/ to remain simple enough to be actually applied to a large number of cases without too heavy a computational effort.
2°/ to provide a simple physical picture of the process.
3°/ to be applicable to a wide class of molecules including closed and open shell ones, diatomics and polyatomics.
4°/ to be fully 'ab initio', without any adjustable parameter, so as to be used as a tool for the interpretation of experimental data.

B. METHOD

The method applied in Ref.(2) to fixed nuclei elastic scattering by a series of closed shell diatomics meets at once the first two preceeding requirements.

1°) Simplifications of the traditional formalism.

It describes the e-molecule interaction through an effective local potential $V(\vec{r})$, using the exact SCF static potential, but the energy dependent Hara approximation for exchange[3] and a model polarisation potential involving a cut-off radius r_c:

(1) $$V_{pol}(\vec{r}) = -(\alpha_0 + \alpha_2 \ P_2 \ (\hat{r}) \ ) \ \{ 1 - \exp - ((\frac{r}{r_c})^6)\}/2r^4$$

Within each irreducible representation λ of the symmetry group of the molecule, the scat- tering equation:

(2) $$\left[-\frac{1}{2r} \frac{\partial^2}{\partial r^2} r + \frac{\vec{\ell}^2}{2r^2} + V(\vec{r}) \right] \psi^\lambda (\vec{r}) = E \ \psi^\lambda (\vec{r})$$

is then solved replacing the usual spherical harmonics basis $Y_\ell^\lambda (\hat{r})$, convenient for a central field problem, by an 'adiabatic angular basis' $\phi_\ell^\lambda (r;\hat{r})$ adapted to the ac- tual anisotropy of the potential and defined so as to diagonalize the fixed-r hamil- tonian:

(3) $$\left[\frac{\vec{\ell}^2}{2r^2} + V(\vec{r}) - \epsilon_\ell^\lambda (r) \right] \phi_\ell^\lambda (r;\hat{r}) = 0$$

The $\phi_\ell^\lambda (r;\hat{r})$ tend to the $Y_\ell^\lambda (\hat{r})$ as soon as anisotropy is negligible that is:
a) at very small r,

b) at very large r,

c) for large ℓ, where $\dfrac{\vec{\ell}^2}{2r^2} \gg V(\vec{r})$.

The ℓ index that labels the different solutions $\phi_\ell^\lambda (r;\hat{r})$ reduces then to the exact orbital quantum number. Expansion of the continuum orbital $\psi^\lambda (\vec{r})$ on this basis:

$$(4) \qquad \psi^\lambda (\vec{r}) = \sum_\ell \frac{1}{r} F_\ell^\lambda (r) \phi_\ell^\lambda (r;\hat{r})$$

leads to the following radial equations

$$(5) \qquad \left[-\frac{1}{2} \frac{d^2}{dr^2} + \varepsilon_\ell^\lambda (r) - E \right] F_\ell^\lambda (r) = \frac{1}{2} \sum_{\ell'} \left[Q_{\ell\ell'}^\lambda (r) + 2 P_{\ell\ell'}^\lambda (r) \frac{d}{dr} \right] F_{\ell'}^\lambda (r)$$

where the non adiabatic couplings

$$(6a) \qquad P_{\ell\ell'}^\lambda (r) = \left\langle \phi_\ell^\lambda (r;\hat{r}) \frac{\partial}{\partial r} \phi_{\ell'}^\lambda (r;\hat{r}) \right\rangle_{\hat{r}}$$

$$(6b) \qquad Q_{\ell\ell'}^\lambda (r) = \left\langle \phi_\ell^\lambda (r;\hat{r}) \frac{\partial^2}{\partial r^2} \phi_{\ell'}^\lambda (r;\hat{r}) \right\rangle_{\hat{r}}$$

are important only in the vicinity of the nuclear singularities. Moreover, they are important only for a few low-lying channels ℓ', allowing a rapid convergence of the scattering calculation.

2°) Physical transparency of the new approach.

Within the adiabatic approximation that neglects the non adiabatic couplings, the adiabatic energies $\varepsilon_\ell^\lambda (r)$ govern the radial motion of the incoming electron, as the effective potential $V_{eff} (r) = V(r) + \dfrac{\ell(\ell+1)}{2r^2}$ does in a central field problem.
The method then provides a clear physical picture of the process by allowing:
a) a direct visualization of the potential responsible for the trapping of the electron, b) a study of the rôle which the various potential terms play in the trapping mechanism. Illustrative results taken from Ref. (2b) are presented on Fig.1

C. GENERALISATIONS

The present work aims at extending this method to open shell diatomics, thus satisfying part of the third requirement we stated in the introduction.

1°) Open shell scattering equations.

_ . The general case is well illustrated by the example of a π electron coupled to the NO (X $^2\Pi$) ground state to give a $^1\Sigma^+$ scattering state with wavefunction:

$$(7) \quad \Psi(^1\Sigma^+) = \frac{1}{2} \left\{ \left| 2\pi_+^+ c\pi_-^- \right| - \left| 2\pi_+^- c\pi_-^+ \right| + \left| 2\pi_-^+ c\pi_+^- \right| - \left| 2\pi_-^- c\pi_+^+ \right| \right\}$$

where the bound orbitals 1σ 2σ 3σ 4σ 5σ 1π 2π will be later on denoted by:
ϕ_i^c , $i = 1...7$ for the closed shells, ϕ_i^o $i = 1,2$ for the open shells, the continuum orbital $c\pi$ by f, the lower (upper) index $\pm$ refers to the sign of the L_z (S_z) component, and the closed shells have been omitted from the notation of the Slater determinants for simplicity. The Schrödinger equation $(H - E) \Psi (^1\Sigma^+) = 0$, projected on the target ground state, leads to the static exchange equation:

$$(8) \quad (h + V^0 - \frac{1}{2} k^2) c\pi_- + V^2 c\pi_+ = W_{exc}^{biel} c\pi_- + W_{exc}^{non\perp} c\pi_-$$

where $\dfrac{k^2}{2}$ denotes the energy of the incoming electron, h the monoelectronic operator

including the kinetic energy and the interaction with the fixed nuclei.
-. The direct potential on the LHS contains two local terms V^0 and V^2.

$$(9) \qquad V^0(\vec{r}) = \sum_i \; 2 < \phi_i^c \mid \frac{1}{|\vec{r}-\vec{r}'|} \mid \phi_i^c > + < 2\pi \mid \frac{1}{|\vec{r}-\vec{r}'|} \mid 2\pi >$$

is the usual static potential depending only on r and θ. It comes from the diagonal matrix elements of the e-target interaction hamiltonian between the unperturbed target states, whereas

$$(10) \qquad V^2(\vec{r}) = \; < 2\pi_+ \mid \frac{1}{|\vec{r}-\vec{r}'|} \mid 2\pi_- >$$

comes from the off diagonal ones. It depends on r,θ and ϕ and arises only in the case of an electron of symmetry $\lambda \neq \sigma$ approaching an open shell molecule in the same symmetry state $\Lambda \neq \Sigma$
-. The exchange contribution on the RHS contains two non local terms W_{exc}^{biel} and $W_{exc}^{non\perp}$.

$$(11) \qquad (W_{exc}^{biel} \; c\pi_-)(\vec{r}) = \sum_i \; <\phi_i^c \mid \frac{1}{|\vec{r}-\vec{r}'|} \mid c\pi -> \phi_i^c \, (\vec{r}) - <2\pi_+ \mid \frac{1}{|\vec{r}-\vec{r}'|} \mid c\pi -> 2\pi_+(\vec{r})$$

$$- <2\pi_- \mid \frac{1}{|\vec{r}-\vec{r}'|} \mid c\pi -> 2\pi_- \, (\vec{r})$$

is the usual bielectronic exchange term, whereas

$$(12) \qquad (W_{exc}^{non\perp} \; c\pi_-)(\vec{r}) = <1\pi \mid c\pi > \{ (h + V^0 - \frac{1}{2} \, k^2) 1\pi + V^2 \, 1\pi_+ - W_{exc}^{biel} \quad 1\pi_- \}$$

$$- <2\pi \mid c\pi > \{ (h + \sum_i \; 2 <\phi_i^c \mid \frac{1}{|\vec{r}-\vec{r}'|} \mid \phi_i^c > - \frac{1}{2} \, k^2) 2\pi - \sum_i <\phi_i^c \mid \frac{1}{|\vec{r}-\vec{r}'|} \mid 2\pi > \phi_i^c \}$$

comes from the non orthogonality of the continuum to the bound orbitals.
-. The coefficients of the integrals in V^0 are orbital occupation numbers. The coefficients of the closed shell terms in W_{exc}^{biel} are all unity. The remaining coefficients in W_{exc}^{biel} and V^2 can be found in the tables of Ref.(4). In the case of a closed shell target, the non orthogonality exchange term reduces to the well known form

$$(13) \qquad \sum_i \; < \phi_i^c \mid f > (\varepsilon_i - \frac{1}{2} \, k^2) \; \phi_i^c$$

where ε_i is the monoelectronic energy of the ith orbital, but no comparable simplification occurs in the open shell case. The only thing that can be shown easily is that the closed shell part of $W_{exc}^{non\perp}$ f reduces to

$$(14) \qquad \sum_i \; < \phi_i^c \mid f > G \, \phi_i^c$$

where G contains all operators of the Fock equation except $W_{exc}^{non\perp}$. One can check that the non orthogonality term in (12) and (13) have the expected form given in (14). This simplification comes from the fact that the total wavefunction is independent of the components of f on the closed shells ϕ_i^c. This latter fact is usually exploited to set the closed shell contribution of $W_{exc}^{non\perp}$ f to zero by choosing f to be orthogonal to the ϕ_i^c's.

2°) New direct term V^2 and local exchange in the open shell case.

-. The treatment of open shell molecules requires then the evaluation of the new direct term V^2 (r, θ, ϕ) that can be added to the main V^0 $(r\theta)$ one, after the angular integration(5).
-. The Hara Free Electron Gaz approximation introduced to deal with the exchange terms in closed shell molecules needs some extensions that we will illustrate in the case of (NO $X^2\Pi + c\pi)^1\Sigma^+$ again. The incoming electron is described by a plane wave of wavenumber $\vec{\kappa}$, the molecule by a set of free electrons enclosed in a box of volume $V = L^3$. A bound electron is then described by a plane wave with quantized wavenumber $k_{x,y,z} = 2\pi n_{x,y,z}/L$ ($n_{x,y,z}$ integers) associated with a given spin state. Accordingly, each cell of volume $8\pi^3/V$ surrounding a given point in momentum space can contain two bound electrons. Hence, the set of closed shell electrons is

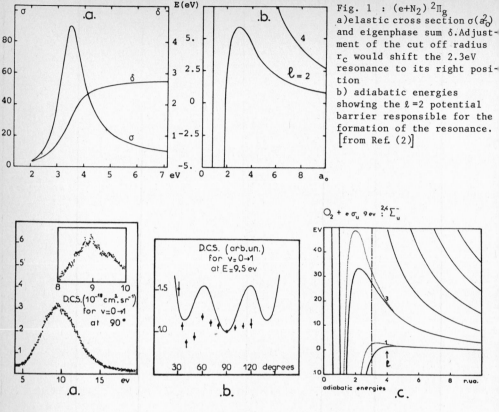

Fig. 1 : $(e+N_2)$ $^2\Pi_g$
a) elastic cross section $\sigma(a_o^2)$ and eigenphase sum δ. Adjustment of the cut off radius r_c would shift the 2.3eV resonance to its right position
b) adiabatic energies showing the $\ell=2$ potential barrier responsible for the formation of the resonance. [from Ref. (2)]

Fig. 2 : O_2. Experimental results from Ref. (1). b) ♦ experimental points; —— theoretical curve for a pure fσ wave. c) the vertical dashed line limits the 'non adiabatic' region; ······ doublet results, —— quartet results, which can be separated only in the $\ell = 1,3$ channels

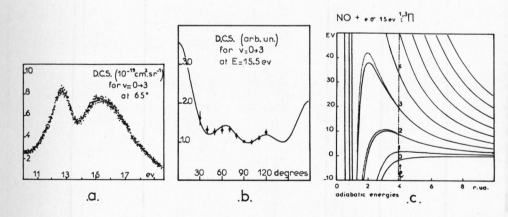

Fig. 3 : NO. Experimental results from Ref. (1). b) ♦ experimental results; —— theoretical curve for a dσ-fσ mixing. c) the vertical dashed line limits the 'non adiabatic' region. ······ singlet results; ——triplet results, which can be separated only in the $\ell = 2,3$ channels.

described in momentum space by a sphere Σ_c whose radius k_c is determined by :

(15) $\qquad \frac{4}{3} \pi k_c^3 = \frac{8\pi^3}{V} \times \frac{14}{2} \quad$ or $\quad k_c^3 = 3\pi^2 \rho_c$

whereas the open shell electron corresponds to a shell Σ_o of internal and external radii k_c and k_F, the latter being determined by:

(16) $\qquad \frac{4}{3}\pi k_F^3 = \frac{8\pi^3}{V} \left(\frac{14}{2} + 1 \right) \quad$ or $\quad k_F^3 = k_c^3 + 6\pi^2 \rho_o$

where we have introduced the densities of closed and open shells electrons ρ_c and ρ_o. Enforcing its local character, and replacing all orbitals by plane waves, the bielectronic exchange term may be rewritten:

(17) $\qquad W_{exc}^{biel} (\vec{r}) = \int_{\Sigma_c} \vec{dk} \times \left[\frac{V}{8\pi^3} \times \vec{dr}' \frac{1}{V} \frac{e^{i(\vec{\kappa}-\vec{k})(\vec{r}'-\vec{r})}}{|\vec{r}-\vec{r}'|} \right] - 2\int_{\Sigma_o} \vec{dk} \times \left[\text{same integrand} \right]$

where the factor 2 comes from the presence of two equal exchange terms for the open shell. Both integrations in equation (17) are then carried out following Slater[6] and give

(18) $\qquad W_{exc}^{biel} (\vec{r}) = \frac{2}{\pi} \left[3k_c F\left(\frac{\kappa}{k_c}\right) - 2k_F F\left(\frac{\kappa}{k_F}\right) \right]$

where F reads

(19) $\qquad F(x) = \frac{1}{2} + \frac{1-x^2}{4x} \ln \left| \frac{1+x}{1-x} \right|$

According to Hara's prescriptions[3], κ is then defined in terms of the Fermi momentum k_F, the target ionization energy I, and the continuum electron energy E by

(20) $\qquad \kappa^2 = k_F^2 + 2 (E+I)$

and the electronic densities ρ_c and ρ_o are taken to be the local ones calculated from the SCF wavefunction of the target.
It is easily seen that the non-orthogonality exchange term vanishes within the above approximation because of the orthogonality of distinct plane waves. The model we present here in a particular case can be immediately extented to any case of open shell configuration.

3°) Results for the open shell diatomics NO and O$_2$

Adiabatic energies and angular eigenmodes have been obtained for both open shell diatomics NO and O$_2$ and for all symmetries of the incoming electron up to δ. Many channels $\varepsilon_\ell^\lambda (r)$ exhibit potential barriers that could be responsible for the existence of shape resonances, but most of them in fact can be excluded, by arguments at the adiabatic level or extrapolation beyond this approximation: thus, the sensitivity of the results to the negative ion state, to the incident electron energy through the exchange potential, to the cut-off radius r_c of the polarisation potential, have been tested. The tunneling probability of an electron of given energy through a given barrier has been estimated from the height and width of the latter and related to the lifetime of the experimentally observed resonance. The r-range of the non adiabatic couplings has been evaluated from plots of the angular eigenmode components on a spherical harmonics basis. Their strength has been roughly estimated from the closeness of the various adiabatic energies $\varepsilon_\ell^\lambda (r)$ within the non adiabatic zone. The interpretation of our preliminary results leads us then to extract the channels that may be responsible for the formation and decay of the observed shape resonances according to:

	low energy resonances	high energy resonances
NO	$\pi \quad \ell = 1,2$	$\sigma \quad \ell = 2,3$
O$_2$	$\pi_g \quad \ell = 2$	$\sigma_u \quad \ell = 3$

the δ (or δu, δg) channels being always repulsive. Both cases would then be consistent with a 'chemical interpretation' relating shape resonances to the trapping of the incident electron in the unoccupied valence orbitals of the target. Results concerning the high energy resonances in NO and O$_2$ are displayed on Fig. 2-3. They agree with bound state calculations for the negative O_2^- ion [7]. They are also consistent with photoionization studies which predict the $\sigma(\sigma u)$ nature of the high lying resonances using the CMSM[8] or the static exchange (S.E.) approximation [9]. On the other hand, the S.E calculations of Ref.(10) using the Tchebycheff imaging technique, show a resonant behaviour in other channels as well.

D. CONCLUSIONS

The present work allows us to visualize the essential features of the e-molecule interaction in the case of the open shell diatomics NO and O_2 in their ground states. It suggests the interpretation of the structures observed in vibrational excitation cross sections at high energy (15eV, 9eV) in terms of $\sigma(\sigma u)$ shape resonances. However, this hypothesis raises further questions, in particular about the fact that both spin states would then seem to be separated in NO ($^{1,3}\Pi$) but not in O_2 ($^{2,4}\Sigma_u^-$). Its confirmation is to be expected only from the completion of this work, yielding accurate eigenphase sums and cross sections. We also plan to test the extended Hara approximation against an exact treatment for exchange, and to introduce the parameter free correlation-polarisation potential of Ref.(11) in our formulation.

REFERENCES

.1. 'Absolute differential cross sections for excitation by electron impact of the v = 1 level in diatomic molecules N_2, CO, NO, O_2.'
    M. Tronc, Y. Le Coat, R. Azria, L. Malegat : to be published.

.2. a) M. Le Dourneuf, Vo Ky Lan, J.M. Launay, J.Phys.B 15 (1982) L 685
    b) Vo Ky Lan, M. Le Dourneuf, J.M. Launay, in 'e-atom and e-molecule collisions'
    (Ed. J. Hinze), Proceedings of the Bielefeld International Workshop, Plenum Press (1983) P. 161

.3. S. Hara, J.Phys.Soc.Japan 22 (1967) 710

.4. P.S. Bagus, Documentation for Alchemy R.J.1077 (# 17873)

.5. G. Raseev, Comp.Phys.Comm.20 (1980) 267

.6. J.C. Slater, 'Quantum theory of molecules and solids', Vol II, App 22, Mac Graw Hill 1963

.7. G.Das, A.C.Wahl, W.T.Zemke, W.C.Stwalley, J.Chem.Phys. 68 (1978) 4252

.8. P.M.Dittman, D.Dill, J.L.Dehmer, J.Chem.Phys. 76 (1982) 5703

.9. G.Raseev, H.Lefebvre-Brion, H. Le Rouzo, A.L.Roche, J.Chem.Phys. 74 (1981) 6686

.10.A.Gerwer, C.Asaro, B.V.Mc.Koy, P.W. Langhoff, J. Chem. Phys. 72 (1980) 713

.11. N.T.Padial, D.W.Norcross, same volume.
    J.K. O'Connell, N.F. Lane, Phys Rev A 27 (1983) 1893.

CONTINUUM MULTIPLE SCATTERING Xα CALCULATIONS OF

ELASTIC CROSS SECTIONS IN ELECTRON—MOLECULE SCATTERING

J. E. Bloor
Chemistry Department
University of Tennessee

Knoxville, TN 37996-1600/USA

Introduction

In recent years several research groups have initiated experimental programs using electron transmission spectroscopy (ETS) to probe the existence of low lying negative ion states in polyatomic molecules.[1-8] Since in many cases more than one negative ion state is revealed by the observation of multiple resonances in the ETS experiment[7,8], any acceptable theoretical model for the interpretation of these results must be free of any parameterization which is dependent on a prior knowledge of the nature of the resonances. Techniques, which depend on parameter adjustment until the first calculated resonance is in agreement with the experimentally observed one, are of little predictive value for molecules such as the chloroethylenes which contain multiple resonances.[7] At the present time two models which avoid this dilemma are available.[9-14] The first one, which is primarily a one center model, is the method recently developed by Jain and Thompson (JT).[9,10] It has proved successful in providing a detailed interpretation of electron scattering results for methane, water, and ammonia and is discussed elsewhere in this volume. The second model is the continuum multiple scattering Xα (MSXα) method originally proposed by Dehmer and Dill[11],but which we have modified so that it is a practical non-empirical method capable of producing total elastic cross sections and differential cross sections for polyatomic molecules containing up to twenty atoms and for electron impact energies from 0 to 100 eV.[12-14] The principle differences between our model and their original method[11] have been summarized elsewhere (Table I of ref. 14).

Although the primary value of the MSXα method is as a survey tool to aid in classifying resonances in polyatomic molecules which are too large to be treated by more accurate methods, it is helpful in assessing the validity of the method to compare it's results on small molecules with results obtained by more accurate methods. For example we have demonstrated[14,15] that the total integral and differential cross sections for elastic scattering of nitrogen over the 0 - 100 eV range are very similar to those obtained by the use of a potential generated from an extended basis set Hartree Fock calculation combined with a semi-empirical polarization potential.[16]

Very recently Abusalbi et al.[17] have used a similar high quality potential to calculate the elastic scattering and rotational excitation cross sections for methane at an impact energy of 10 eV. We compare in the next section our MSXα calculations with this result, with the earlier work of Gianturco and Thompson (GT)[18], with the later JT result[9] which uses a non-empirical polarization (AT) potential, and with recent experimental observations. In the final section we describe MSXα results for furan and thiophene as examples of our ongoing research program comparing molecules containing elements of the second row of the periodic table with their third row analogues.

Methane Cross Sections

The availability of a number of recent experimental studies[19-21] on the total electron scattering cross sections, momentum transfer cross sections, and differential cross sections (DCS) in the impact energy range 0 - 20 eV make this molecule a prime target for the evaluation of theoretical calculations. For example GT[18] have used this molecule in testing a number of different models, culminating in the recent JT[9,10] parameter free model. The results of our MSXα calculation for total elastic cross sections are compared with experiment and with the recent JT results[9] in Fig. 1a. It can be seen that in the region from 1 - 15 eV the MSXα and the JT results are very similar and are both consistently higher than recent experimental results.[19,20] Both models reproduce the Ramsauer-Townsend minimum below 1 eV, although the JT result is closer to the recent measurements of Barbarito et al.[20] than is the MSXα method, which on the other hand agrees better with the older results of Bruche and of Ramsauer and Kollath shown in ref. 20. Fig. 1b shows that the momentum transfer cross section calculated by the MSXα method is in relatively good agreement with experiment[19,21] over the energy range 0.10 to 20 eV.

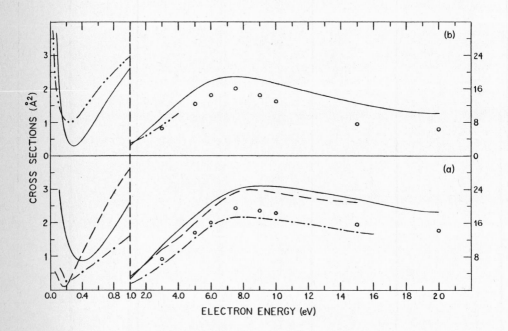

Figure 1

Methane total cross sections from 0.10 to 20.0 eV. Broken vertical line denotes change of scale at 1.0 eV. a) Integral cross sections. Theoretical:———, this work;— — —, JT[9] with AT potential. Experimental:—·—, Barbarito et al[20]; ○, Tanaka et al[19]. b) Momentum transfer cross sections. Theoretical:———, this work. Experimental:—··—, Duncan and Walker[21]; ○ , Tanaka et al[19].

It is well known that the DCS is a much more sensitive probe of model potentials than is the total cross section. As will be described fully elsewhere,[15] our MSXα calculations on the DCS of nitrogen (0 - 100 eV) and methane (0 - 20 eV) are in good agreement with experiment. This is demonstrated in Fig. 2 for an impact energy of 10 eV. Unfortunately theoretical results using the JT model are not available at this energy. However, we have compared our results with the previous best results of GT using an empirical cut-off potential[18]; with results generated by Abusalbi et al.[17] using a Hartree Fock ab initio potential; and with the experimental results of Tanaka et al.[19] In general all three calculations agree with the experiments. In the 60° - 80° range the MSXα calculation appears to exaggerate the percentage d character in the resonant T_2 channel by about 5% with the d wave comprising 64% of the total cross section for all channels at this energy. Apart from this slight discrepancy the partial wave contributions agree very closely with those reported by Abusalbi et al.[17] It should be noted that the broad resonance in the 5 - 9 eV region is due mainly to the T_2 channel which never exceeds more than 65% d wave character in our calculation, contrary to Tanaka's suggestion[19] that this resonance is dominated by d wave scattering.

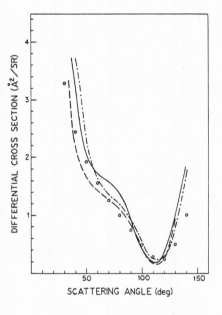

SCATTERING ANGLE (deg)

Figure 2

Methane differential cross sections at 10.0 eV. Theoretical: ———, this work; — — —, GT[18] with cut-off parameter r_o = 0.92; — . —, Abusalbi et al[17] with SEPlke potental and Basis III. Experimental: ○, Tanaka et al.[19]

Furan and Thiophene Cross Sections

The understanding of the difference between elements of the second and third rows of the periodic table when substituted into conjugated systems is a matter of great contemporary interest.[4] Although a great deal of experimental information on the effect of such substituents on occupied energy levels has been amassed over the last decade using photoelectron spectroscopy, the collection of similar data on the unoccupied levels, using ETS, is in its infancy.[1-4,7]

One of the first applications of ETS to this problem was a study of the effect on the unoccupied energy levels of ethylene by replacing successive hydrogens with fluorine and/or chlorine.[7] More recently a number of studies on the oxygen and sulfur containing compounds have been reported.[2-4] The results on fluorine and chlorine substitution were very unexpected and propose a severe test for any electron-molecule scattering model. Fluorine substitution was found, contrary to qualitative arguments and to the results of Hartree Fock calculations on the neutral molecules, to destabilize the level of ethylene. In addition it was experimentally established that, whereas in the fluoroethylenes only π* resonances were observed, two types of resonances were found to be present in the chloroethylenes. The MSXα method has been shown to give a satisfactory theoretical interpretation of these experiments.[13]

Calculations on other molecules containing second and third row elements are in progress[15]. We present in Table 1 preliminary results on furan and thiophene.

Table 1

MSXα calculated and observed resonance positions for shape resonances in furan and thiophene below 10 eV.

| Type of | Furan | | Thiophene | |
resonance	MSXα (eV)	Exp. ETS (eV)	MSXα (eV)	Exp. ETS (eV)
b_2 ($\sigma*$)	7.0	6.6	5.5	5.3
a_2 ($\pi*$)	3.26	3.15	2.68	2.63
b_1 ($\pi*$)	1.63	1.73	1.19	1.15

The experimental results of Modelli et al.[4] clearly show two resonances in both compounds. These are almost certainly due to temporary electron attachment by the two unoccupied π* levels of b_1 and a_2 symmetry. The theoretical calculations are in full agreement with this assignment, and indeed the theoretical derivative curve for thiophene shown in Fig. 3 is in very good agreement with the corresponding experimental curve. The theoretical model for thiophene also predicts the presence of a third very broad resonance in the 4.0 – 6.0 eV region due to a σ* (b_2) orbital. Such a very broad resonance is seen in the experimental spectra but Modelli et al. suggest[4] it is due to a two electron process, i.e., a core excited resonance. Modelli et al. also propose that the π* (b_1) resonance in thiophene appears at lower energies than the corresponding resonance in furan because of the participation of 3d orbitals on the sulfur atom in the π* MO. However our MSXα calculations show virtually no sulfur d orbital mixing in either of the two π* orbitals in thiophene.

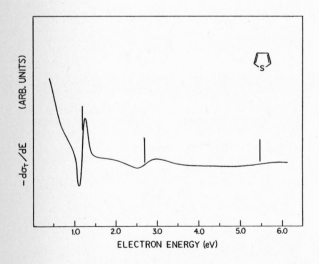

Figure 3

Thiophene theoretical derivative spectrum: Obtained by calculating the change in slope of the total integral cross section over the energy range 0.3 to 6.0 eV. In order of increasing energy, the vertical bars denote respectively the positions of the b_1 (π*), a_2 (π*), and b_2 (σ*) resonances.

In order to ascertain the true reason for the observed differences, an analysis of the nature of the static potential and the magnitude of the distortion caused by the incoming electron is underway.

ACKNOWLEDGEMENTS

This research was supported in part by the National Science Foundation, Grant No CHE 8207040, and by a generous grant of computer time from the University of Tennessee Computing Center. The author is most grateful for the assistance of Mr. R. E. Sherrod in performing the calculations and for many helpful discussions.

References

1   K. D. Jordan and P. D. Burrow: Accts. Chem. Res. 11, 341 (1978).
2   A. Modelli, G. Distefano and D. Jones: Chem. Phys. 73, 395 (1982).
3   A. Modelli, D. Jones, F. P. Colonna and G. Distefano: Chem. Phys. 77, 153 (1983).
4   A. Modelli, G. Distefano, D. Jones and G. Seconi: J. Elec. Spec. Relat. Phenom. 31, 63 (1983).
5. J. C. Giordan, J. H. Moore and J. A. Tossell: J. Am. Chem. Soc. 103, 6632 (1981).
6   J. C. Giordan, J. H. Moore, J. A. Tossell and J. Weber: J. Am. Chem. Soc. 105, 3431 (1983).
7   P. D. Burrow, A. Modelli, N. S. Chiu and K. D. Jordan: Chem. Phys. Lett. 82, 270 (1981).
8   P. D. Burrow, A. Modelli, N. S. Chiu and K. D. Jordan: J. Chem. Phys. 77, 2699 (1982).
9   A. Jain and D. G. Thompson: J. Phys. B, At. Mol. Phys. 15, L631 (1982).
10 A. Jain and D. G. Thompson: J. Phys. B, At. Mol. Phys. 16, L347, 1113, 2593 (1983).
11 D. Dill and J. L. Dehmer: in Electron-Molecule and Photon-Molecule Collisions (eds. T. Resigno, V. McKoy and B. Schneider). Plenum Press, New York 1979.
12 J. E. Bloor, R. E. Sherrod and F. A. Grimm: Chem. Phys. Lett. 78, 351 (1981).
13 J. E. Bloor and R. E. Sherrod: Chem. Phys. Lett. 88, 389 (1982).
14 J. E. Bloor: Paper presented at the David Bates Symposium in March 1983. To be published in Int. J. Quant. Chem. Symp. (1983).
15 J. E. Bloor and R. E. Sherrod: In preparation (1983).
16 J. R. Rumble, Jr., D. G. Truhlar and M. A. Morrison, J. Chem. Phys. 79, 1846 (1983).
17 N. Abusalbi, R. A. Eades, T. Nam, D. Thirumalai, D. A. Dixon and D. G. Truhlar: J. Chem. Phys. 78, 1213 (1983).
18 F. A. Gianturco and D. G. Thompson: in Electron-Atom and Electron-Molecule Collisions (ed. J. Hinze). Plenum Press, New York 1983.
19 H. Tanaka, T. Okada, L. Boesten, T. Suzuki, T. Yamamoto and M. Kubo: J. Phys. B, At. Mol. Phys. 15, 3305 (1982) and references therein.
20 E. Barbarito, M. Basta and M. Calicchio: J. Chem. Phys. 71, 54 (1979).
21 C. W. Duncan and I. C. Walker: J. Chem. Soc. Faraday Trans. II 68, 1514 (1972).

PROJECTION OPERATOR CALCULATIONS FOR MOLECULAR SHAPE RESONANCES IN THE OPTICAL POTENTIAL APPROACH

Michael Berman and Wolfgang Domcke,
Theoretische Chemie, Physikalisch-Chemisches Institut,
Universität Heidelberg, D-69 Heidelberg, West-Germany

Despite its usefulness as a qualitative tool, the projection operator formalism of Feshbach /1/ has found very little application as a quantitative computational tool for shape resonances. A noteable exception are the recent calculations of Hazi and coworkers on molecular shape resonances /2/.

In this article we present results obtained with a new method to decompose the fixed-nuclei T matrix for electron-scattering (or, equivalently, the eigenphase sum) into a resonant and a background contribution, such that the background term becomes a smooth function of E and R, while rapid variations of the eigenphase sum with E and R are contained in the resonant term. The first step is to reduce the electronic many-body problem to an effective one-body scattering problem using the many-body optical potential formalism /3-5/. The resulting energy-dependent, non-local and in general complex optical potential is given by the irreducible self-energy part of the many-body Green's function formalism /3-5/. We then apply projection operator techniques /1/ to this effective one-body problem to extract rapid variations of the phase shift due to shape resonances or virtual states. This requires the construction of a suitable (in general energy dependent) discrete state, using, for example, the stabilization method /6,7/. In contrast to the approach of Hazi /2/, the formalism yields not only the width and level-shift functions needed for the treatment of nuclear dynamics in the resonance state, but also the background T matrix or eigenphase sum. Moreover, because the background scattering problem is solved with proper boundary conditions, we obtain the information on the angular distribution of the resonant and background scattering. The two-particle-hole Tamm-Dancoff approximation (2ph-TDA) /8,9/ is adopted for the optical potential and the Schwinger variational principle /10/ is used to solve the background scattering problem. To test the performance of the method, we have applied it to the 2.3 eV $^2\Pi_g$ shape resonance in N_2, which is by far the best studied molecular shape resonance, both experimentally and theoretically. In particular, we shall compare our results with the data obtained by Hazi /2/ using a rather different computational approach. The excellent agreement of the present results with the results of Hazi and experimental data indicate that the systematic ab initio calculation of resonance positions and widths to an accuracy of 0.1 eV or better is indeed feasible.

In the results to be presented we confine ourselves to fixed-nuclei calculations at the equilibrium geometry R_o = 2.068 a.u.. The separation of the $^2\Pi_g$ eigenphase sum into background (dashed-dotted) and resonance (dashed) contributions in the 2ph-TDA is shown in Fig. 1. We have checked (numerically) that the background and resonant eigenphase sums add up to the full eigenphase sum (full line in Fi.g 1a) calculated directly from the optical potential as in ref. 9.

Fig. 1 shows that the background contribution is relatively small and weakly energy-dependent. The resonant contribution shows the typical resonance behaviour, i.e. a rapid increase in the eigenphase sum near 2.3 eV. The resonance position is close to the experimental value /11/.

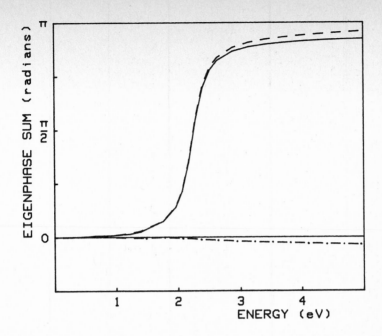

Fig. 1: Eigenphase sum of the $^2\Pi_g$ resonance in e-N$_2$ scattering in the 2ph-TDA. Full line denotes the full eigenphase sum, the dashed line denotes resonance eigenphase sum and the dashed-dotted denotes the background contribution.

The corresponding width and level-shift functions are shown in Fig. 2 (full lines). For comparison, the width and level-shift functions calculated by Hazi /2/ are included in Fig. 2 (dashed lines). The agreement between the two width functions (in particular near the resonance energy) is excellent, considering that the two calculations are performed with completely different methods and that different basis sets are used.

In Fig. 3 we compare the resonant eigenphase sum of the present work (full line) with the resonant eigenphase sum calculated from Γ and Δ given by Hazi /2/ (stars). It is seen that the two result are in excellent agreement. In particular, the resonance positions agree to an accuracy of 0.1 eV. The widths at the resonant position are also in excellent agreement (see Fig. 2).

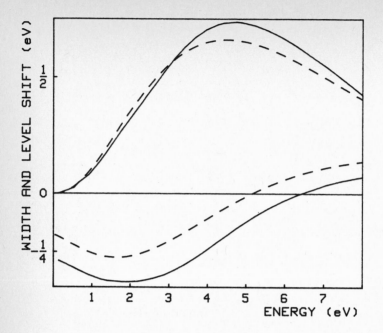

<u>Fig. 2:</u> b) Width and level-shift functions in the 2ph-TDA as a function of
energy. Full lines denote results of present work, while the dashed
lines stand for the results of Hazi /2/.

In conclusion, the main advantages of the present approach /12/ are the
following: (i) In addition to the resonant T-matrix, width and level shift, we
obtain the background scattering T-matrix. (ii) We obtain explicitly the
information on the angular distribution of the resonant and background
scattering. (iii) Target polarization and correlation are included in a
balanced way via the optical potential.

References

1. H.Feshbach, Ann. Phys. (N.Y.) <u>19</u>, 287 (1962)
2. A.U. Hazi, J. Phys. B <u>11</u>, L259 (1978);
   A.U. Hazi, in <u>Electron-Molecule and Photon-Molecule Collisions</u>,
   edited by T.N. Rescigno, V. McKoy and B.I. Schneider (Plenum
   Press, New York, 1979), p. 281;
   A.U. Hazi, in <u>Electron-Atom and Electron-Molecule Collisions</u>,
   edited by Hinze (Plenum Press, New York, 1983) p. 103
3. J.S. Bell and E.J. Squires, Phys. Rev. Lett. <u>3</u>, 96 (1959)

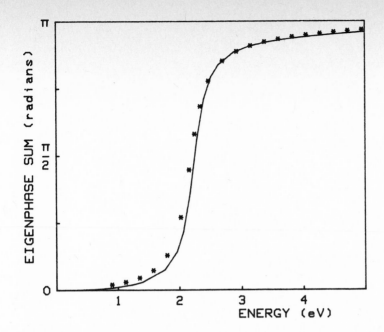

Fig. 3: Resonant $^2\Pi_g$ eigenphase sum of e-N_2 scattering calculated in this
work (full line) compared to the results of Hazi /2/ (stars).

4. M. Namiki, Progr. Theor. Phys. 23, 629 (1960)
5. J.S. Bell, in Lectures on the Many-Body Problem, edited by E. Caianiello
   (Academic Press, New York, 1962), p. 91
6. E. Holøien and J. Midtdal, J. Chem. Phys. 45, 2209 (1966)
7. A.U. Hazi and H.S. Taylor, Phys. Rev. A 1, 1109 (1970)
8. L.S. Cederbaum and W. Domcke, Adv. Chem. Phys. 36, 205 (1977);
   J. Schirmer and L.S. Cederbaum, J. Phys. B 11, 1889 (1978);
   J. Schirmer, O. Walter and L.S. Cederbaum, Phys. Rev. A 28, 1237 (1983)
9. M. Berman, O. Walter and L.S. Cederbaum, Phys. Rev. Lett. 50 (25), 1979 (1983)
10. R.G. Newton "Scattering Theory of Waves and Particles", second edition
    (Springer-Verlang, N.Y., Heidelberg, Berlin, 1982) p. 319;
    R. Sugar and R. Blankenbecler, Phys. Rev. 136, B472 (1964);
    S.K. Adhikari and I.H. Sloan, Phys. Rev. C 11, 1133 (1975);
    D.K. Watson and V. McKoy, Phys. Rev. A 20, 1474 (1979);
    D.W. Watson, R.R. Lucchese, V. McKoy and T.N. Rescigno, Phys. Rev. A 21,
    738 (1980);
    K. Takatsuka and V. McKoy, Phys. Rev. A 23, 2352 (1981);
    L. Mu-Tao, K. Takatsuka and V. McKoy, J. Phys. B 14, 4115 (1981);
    K. Takatsuka and V. McKoy, Phys. Rev. A 24, 2473 (1981)
11. G.J. Schulz, Rev. Mod. Phys. 45, 423 (1973)
12. For more details we refer the reader to the forthcoming papers,
    W. Domcke, Phys. Rev. A, in press
    M. Berman and W. Domcke, Phys. Rev. A, submitted.

ELECTRON SCATTERING BY N_2:

FROM MUFFIN-TIN TO MCSCF POTENTIALS

John R. Rumble, Jr.
National Bureau of Standards
Washington, DC   20234

Introduction

Over the past few years, we have performed calculations for the scattering of electrons by N_2[1] in which the target electron density is described by several wavefunctions which differ both in accuracy and ease of computation. The aim of these calculations has been twofold: First, to provide a set of well-converged results for scattering by N_2 of intermediate energy electrons; second, to determine the accuracy of the various wavefunction models using a consistent scattering formulation. The hope has been to calibrate the models so that in similar calculations for larger molecular systems, in which it will not be feasible to use even Hartree-Fock wavefunctions, less accurate models can be used with an understanding of the limitations.

Four methods of obtaining the electron density of the target N_2 have been used:

    1) INDO
    2) Muffin-Tin (MT)
    3) Hartree-Fock (HF)
    4) MCSCF-FOCI (CI)

In all cases, once the density matrix for each wavefunction has been found, the same computer codes have been used to generate

    1) the Static Potential, V_S,
    2) the Exchange Potential, V_E, and
    3) the Polarization Potential, V_P

The total scattering potential was simply

$$V_{TOT} = V_S + V_E + V_P$$

The static potential was calculated exactly in terms of its Legendre components. The exchange potential was calculated using two local exchange approximations, namely, the Hara-Free-Electron-Gas Exchange (HFEGE) Model and the Semiclassical Exchange (SCE) Model. The polarization was modeled by a parameterized function using a cutoff radius determined by matching calculation to experimental results at the 2.39 eV Π_g resonance. The scattering problem was treated in a body-fixed reference frame using the fixed-nucleus approximation at various internuclear distances. To integrate the coupled channel equations, the integral equation algorithm was used. Results were converged to about 5 percent with respect to the step size, number of channels, and number of potential terms.

State-to-state differential, integral, and momentum transfer cross-sections were calculated from transition matrix elements by a computer program of Henry.[2] The transition matrix elements were calculated from body-frame reactance matrix elements as taken from the close-coupling solution for $m \leq 3$ and from the unitarized polarized Born approximation for $m > 3$.

Charge Density Description Tested

A. <u>INDO</u>

This is a semiempirical method developed for calculating the geometry of complicated organic molecules which cannot be treated <u>ab initio</u>. The valence electrons
are treated explicitly; however, the two-center electron repulsion integrals are
neglected. The resulting wavefunctions usually predict geometrics fairly well
but energies are less accurate. The advantage that the method has is the greater
ease of calculating electron densities over a wide range of nuclear geometrics by
electron impact.

B. <u>Muffin-Tin</u>

A second semiempirical model of a chemically bonded system is the so-called
"muffin-tin." Here the electrons of the molecule are assigned to one of two
distinct volumes surrounding the nuclei. The innermost regions are atomic-like
spheres around each nuclei. Next, the chemical bonding electrons are confined
to a sphere (region II) which surrounds the molecule as a whole. Usually the
atomic spheres touch and are inscribed inside the density. The appeal of this
model is its simplicity, both in physical understanding and in calculations.
For our calculations, we used the HF electron densities in the atomic sphere and
a uniform electron density in region II. The HFEGE exchange model was also used
here. Only polarization effects are included outside region III.

The major drawback of this model is the discontinuity in electron density and
therefore the potential at the point the spheres meet. While smoothing functions
could be used, these complicate the model and reduce its appeal.

C. <u>Hartree-Fock</u>

Converged Hartree-Fock wavefunction of Cade, Sales, and Wahl[3] were used extensively. A good basis set was used, and both the electron density and energies
compare very well with even more converged calculations.[4]

D. <u>MCSCF-FOCI</u>

The wavefunction calculations were carried out with the ALCHEMY system of codes.
A full valence multiconfiguration SCF calculation was first carried out within
the basis set described elsewhere.[5] The predicted dissociation energy at the
experimental R_e = 2.074 bohr is 8.59 eV which is slightly more than 1 eV less
than the experimentally determined dissociation energy of 9.759 eV. A first-
order configuration interaction calculation was then carried out relative to the
18 configuration base wavefunction. The dissociation energy obtained from this
482 configuration calculation was 8.87 eV, which is slightly better than the
MCSCF result. Improvements in the calculation beyond this point were deemed
unnecessary since most of the correlation energy involving the Π_g space is
included in the MCSCF-FOCI wavefunction, and the resulting density matrix should
be nearly optimum.

Results

Some representative results are given below. The major conclusion is that for non-resonant scattering by closed-shell molecules, a Hartree-Fock level target electron density is necessary and sufficient. For resonant scattering, even the Hartree-Fock description is not good enough. Semiempirical descriptions, such as INDO or the Muffin-Tin, appear to be adequate for survey calculations but not good enough for definitive work.

In Table I, we summarize the results for elastic scattering between 5 and 30 eV at the equilibrium internuclear distance using only the static and exchange terms of the potential. The results for the MCSCF-FOCI target are the first beyond-Hartree-Fock results for N_2 within the close-coupling framework.

For non-resonance scattering, all four target wavefunctions give fairly close results, especially as the incident energy increases. For channels which exhibit resonance behavior, very drastic differences show up. Consequently, caution must be used in interpreting results for these channels near resonance energies. For systems which have well-known features such as the 2-4 eV resonance of N_2, the energy ranges and symmetries are available for interpretation. Calculated features not yet seen experimentally will need to be checked thoroughly.

Table I

$e_2 - N_2$

Cross Sections - (a_0^2) - Elastic Scattering

Static + Exchange Potential Only

Channel		Σ_g	Π_g	Σ_u	Π_u	All[c]
Energy	CI[a]	34.18	29.82	14.08	10.05	88.4
5 eV	HF[a]	33.99	57.53	13.18	9.56	114.5
	INDO[b]	34.40	10.51	17.00	8.78	72.2
13.6 eV	CI[a]	11.63	9.75	11.71	12.74	47.4
	HF[a]	11.72	8.68	11.54	12.43	46.1
	INDO[b]	11.87	14.51	12.27	12.28	54.2
	MT[a]	9.35	10.32	11.89	15.02	54.2
30 eV	CI[a]	5.53	6.19	10.45	10.81	36.6
	HF[a]	5.49	5.90	10.64	10.86	36.8
	INDO[b]	6.02	7.24	9.52	10.44	38.1

a - Exchange calculated using HFEGE Model
b - Exchange calculated using SCE Model
c - Sum for $m \leq 4$

In Table II, the resonance parameters, energy, and width are given for the low-energy Π_g resonant channel in N_2 for several different calculations. On the static-exchange level, the resonance energy goes up from 5.06 eV to 5.54 eV. This result is expected since

$$Eres = E\ (N_2^- \ ^2\Pi_g) - E\ (N_2\ ^1\Pi_g).$$

Since the MCSCF-FOCI wavefunction yields a better ground state ($^1\Pi_g$) energy, i.e., lower energy, the energy difference increases.

Table II

Resonance Parameters N_2^- ($^2\Pi_g$)

R=2.068 a_o

Calculation	Polarization Included	E_{RES} (eV)	Γ_{RES} (eV)	Reference
Hartree-Fock	No	5.06	1.66	present
MCSCF-FOCI	No	5.54	1.99	present
Stieltjes	No	4.13	1.14	4
Exact Static-Exchange	No	3.90	1.33	5
Stieltjes	Yes	2.23	0.40	4
R-Matrix	Yes	2.15	0.34	6
Exact Static-Exchange	Yes	2.16	0.47	7

Two separate improvements must be made to the present calculations using the MCSCF-FOCI wavefunction to bring the resonance results in line with the Stieltjes, R-Matrix, and exact results. The first is to improve the scattering calculation itself to account for correlation effects in the N_2^- system. Work is underway to achieve this. The second is to include polarization.

Summary

At the present time, it appears that to calculate the main experimental features seen in the scattering of electrons by N_2 in regions where resonance effects are important, a quite accurate target electron density is required. The non-resonance scattering, however, can be described fairly well in various approximate ways.

References

1. J. R. Rumble, D. G. Truhlar, and M. A. Morrison, J. Chem. Phys. 79, 1846 (1983); J. Phys. B. 14, 2301 (1981); J. Chem. Phys. 70, 4101 (1979); 72, 3441 (E) (1980); J. Chem. Phys. 72, 5223 (1980); J. Chem. Phys. 72, 3206 (1980).

2. R. J. W. Henry, Comput. Phys. Comm. 10, 375 (1975).

3. P. E. Cade, K. D. Sales, and A. C. Wahl, J. Chem. Phys. 44, 1973 (1966).

4. L. C. Collins, W. D. Robb, and M. A. Morrison, Phys. Rev. A. A21, 488 (1980).

5. A. U. Hazi, T. N. Rescigno, and M. Kurila, Phys. Rev. A. A23, 1089 (1981).

6. B. I. Schneider, M. LeDournef, and Vo Ky Lan, Phys. Rev. Letts. 43, 1926 (1979).

7. N. Padial and D. W. Norcross, Phys. Rev. A., in press.

ON THE FINITE VOLUME VARIATIONAL METHOD BASED ON THE LOGARITHMIC DERIVATIVE

OF THE WAVE FUNCTION

G. Raşeev
Laboratoire de Photophysique Moléculaire du CNRS,
Bât. 213, Université de Paris-Sud 91405 Orsay, France
and Département de Chimie, Bât. B6, B-4000 Sart-Tilman,
Liège 1, Belgique.

1.- Introduction

One of the well known and widely used methods to study the collision problems is the R matrix developed in nuclear physics by Wigner and Eisenbud (1) and applied to atomic physics by Burke et al. (2) and to molecular physics by several authors (3-7). This method is based on the expansion of the continuum wave function on several known functions and the separation of the coordinate space into an internal (interaction) and external (asymptotic) regions. The technique of expanding the wave function is widely used in bound calculations and the R matrix borrows some computational techniques from this field. The separation of the coordinate space into two regions is common to all collision methods but in R matrix the size of the internal region is smaller than in other methods. The main advantage of the R matrix is that the functions used in the expansion are calculated only once. The disadvantage is that the R matrix, written as an expansion in terms of these functions at each continuum energy, converges slowly with the number of functions retained in the expansion. To speed this convergence different corrections can be added to the R matrix (7).

Alternatively methods sharing-in part the caracteristics of the R matrix but free from the convergence problems have been developed. First, the eigenchannel method have been used in atomic and molecular physics first by Fano and Lee (8) then by Jungen (9) and Raşeev and Le Rouzo (10) (RLR). It is a stable but cumbersome method. Secondly, a variational version of the R matrix has been introduced by Jackson (11) and further developed by Nesbet (12). This method seems to be able to overcome convergence problems. Finally, an alternative method has been developed for atoms by C. Green (13) and simultaneously for molecules by Le Rouzo and Raseev (14) (LRR). This method, called hereafter Finite Volume Variational Method (FVVM), has been applied preceedingly (LRR) to molecular photoionization of H_2^+ (which is an one electron system).

The FVVM is a straightforward multichannel extension of the logarithmic derivative of the wave function variational method developed in 1948 by Kohn (15). In this method the R matrix is not directly calculated but the diagonalization of the generalized system of eigenequations (Eq. 4 below) gives energy degenerate solutions. Each eigenvalue is variational logarithmic derivative of the wave function at the boundary and the corresponding eigenvectors are expansion coefficients of this same wave function.

In this paper the FVVM is presented in a form suitable for the integration in Electronic Ab-Initio Quantum Defect Theory (EAQDT) (see RLR). We also present the application of FVVM to H_2 photoionization which is a prototype of many electron diatomic molecule. Finally, in concluding the paper we briefly discuss the relative merits of existing molecular finite volume methods.

2.- Résumé of Finite Volume Variational Method (FVVM)

The FVV method is based on the solution of finite volume global Schrödinger equation with a constraint of hermiticity (Eq. 1b) imposed on the hamiltonian matrix

$$< \psi \mid H - E \mid \psi >_V = 0 \qquad (1a)$$

$$< \psi \mid \mathscr{L} (a) \mid \psi >_\Sigma = 0 \qquad (1b)$$

Here V is the finite volume of approximate spherical radius a and surface boundary Σ and $\mathscr{L}$ (a) = $(\partial/\partial n - b)$ $\delta(r-a)$ is the Bloch (21) operator where $\partial/\partial n$ is the normal derivative to the surface Σ and $\delta(r-a)$ is Dirac delta function showing that $\mathscr{L}$ acts only on Σ. b is yet unidentified quantity which will be obtained variationally in solving equation 4 below. Now, we expand the inner region wave function (the wave function confined in V) on a basis of known function χ_i collected in a row vector

$$\underline{\psi}^{in} = \underline{\chi} \ \underline{C} \qquad (2)$$

where $\underline{\psi}^{in}$ is a row vector representing the energy degenerate internal region wave functions. At the boundary Σ we expand the basis functions in surface harmonics $Y_L(r)$, where $L = (\ell,m)$ namely $\underline{\chi}(a) = \frac{1}{a} \ \underline{Y}(\hat{a}) \ \underline{\theta}(a)$, where $\underline{\theta}$ is a matrix each column representing one basis function expansion. The function $\underline{\psi}^{in}$ can be written as

$$\underline{\psi}^{in}(a) = \frac{1}{a} \ \underline{Y}(a) \ \underline{F}(a) \qquad (3)$$

where $\underline{F}(a) = \underline{\theta}(a) \ \underline{C}$. For simplicity, except when explicity stated, we restrict our discussion below to Frozen Core Static Exchange (FCSE) approximation with only one core considered implicitly in the equations. In this case the asymptotic channels are unambiguosly labelled by L the surface harmonics label. We shall have as many energy degenerate solutions labelled α as asymptotic channels. In RLR, we have formulated the eigenchannel method in a more general fully N electron framework. The present method can easily be cast in similar N electron form except that here the basis functions i are independent of the particular energy degenerate solution α. Nevertheless such an attempt will not made here. To proceed further with the derivation, we introduce (1b) and (2) in (1a) and apply the variational principle to the expansion coefficients. We obtain the generalized system of eigenequations

$$\underline{A} \ \underline{C} = \underline{\Delta} \ \underline{C} \ \underline{b} \qquad (4a)$$

where $A_{ij} = <\chi_i \mid H + \partial/\partial n - E \mid \chi_j>_V$ and $\Delta_{ij} = < \chi_i \mid \chi_j >_\Sigma$. The solution of (4a) gives as eigenvalues (diagonal elements of $\underline{b}$) the logarithmic derivative of the wave function at the boundary Σ and as eigenvectors $\underline{C}$, the expansion coefficients of the same function. The number n of basis functions in the expansion (2) which define the order of the matrices $\underline{A}$ and $\underline{\Delta}$ is greater or equal to the number of asymptotic channels n_L. Therefore $n - n_L$ eigenvalues of (4a) should be meaningless. It has been demonstrated (see appendix of LRR paper) that the number of meaningful solutions of (4a) is equal to the number of asymptotic channels n_L. The equation (4a) can now be rewritten in slightly different from namely

$$\underline{A}^{-1} \ \underline{\Delta} \ \underline{C} = \underline{C} \ \underline{b}^{-1} \qquad (4b)$$

As A is nonsingular the matrix $\underline{A}^{-1} \underline{\Delta}$ exists and is nonsymmetric. Two algorithms one due to Moler and Stewart (16) and the other to Wilkinson (17), exist. The merits of the two equations (4) are linked to the stability and efficiency of corresponding algorithms. The algorithm for solving (4b), allows selection of few eigenvalues (n_L in our case) while the algorithm for solving (4a) gives all eigenvalues and eigenvectors. Therefore the algorithm for (4b) seems more efficient. But this algorithm fails if one of the eigenvalues of A is close to zero. Therefore, we can first use (4b) and if it fails then we use (4a).

Now we are in position to obtain the collision matrices and also the wave function over the entire coordinate space. This is done by matching the wave function at the surface Σ to a known asymptotic from $\underline{\mathcal{F}}(r)$ in terms of K matrix

$$\underline{\mathcal{F}}(r) = \sqrt{\frac{2}{\pi k}} \left\{ \underline{f} + \underline{g}\,\underline{K} \right\} \tag{5}$$

where f and g are diagonal matrices with elements $f_\ell(kr) \underset{r \to \infty}{\sim} \sin(kr + \omega_\ell)$ and $g_\ell(kr) \underset{r \to \infty}{\sim} \cos(kr + \omega_\ell)$. The phaseshift ω_ℓ is in the present application centrifugal and Coulomb phaseshift. Now we match the function and the derivative at the boundary and use the eigenvalue matrix b resulting from equations (4) to obtain

$$\underline{F}\,\underline{N} - \underline{g}\,\underline{K} = \underline{f}$$
$$\underline{F}\,\underline{b}\,\underline{N} - \underline{g}'\,\underline{K} = \underline{f}' \tag{6}$$

From (6), we can obtain K and N as

$$\underline{K} = -(\underline{B}\,\underline{g} - \underline{g}')^{-1}\,(\underline{B}\,\underline{f} - \underline{f}') \tag{7a}$$

$$\underline{N} = \underline{F}^{-1}\,(\underline{f} + \underline{g}\underline{K}) \tag{7b}$$

where $\underline{B} = \underline{F}^{-1}\,\underline{b}\,\underline{F}$. Finally, by the diagonalization of (7a), we obtain the quantum defects μ_α and the mixing coefficients $U_{L\alpha}$. Then the energy normalized transition moment for photoionization in terms of K matrix or eigenchannel asymptotic forms is written as

$$\underline{d}^K = \underline{N}\, < \psi \,|\, \vec{r} \,|\, \psi_o > \tag{8a}$$
and
$$\underline{d}^\mu = \underline{\cos \pi \mu}\ \underline{U}^+\ \underline{N} < \psi \,|\, r \,|\, \psi_o > \tag{8b}$$

where d^K and d^μ are column vectors, each element corresponding to an energy degenerate solution of equation (4) ; ψ is the final state wave function defined in (2) for the internal region (no contribution to d from the external region is calculated) and ψ_o is the initial state wave function ; $\underline{\cos \pi \mu}$ is a diagonal matrix, each element being $\cos \pi \mu_\alpha$. The transition moment $\underline{d}^K$ is used in the photoionization cross section expression (see LRR, eqs. 33-35)

$$\underline{d}^{(-)} = [(\underline{1} + i\,\underline{K})^{-1}\,]^+\ \underline{d}^K \tag{9a}$$

$$\sigma = 2.6891\ (I + E)\ (\underline{d}^{(-)})^+\ \underline{d}^{(-)} \tag{9b}$$

In (9b) I is the ionization potential and E the continuum energy. The factor 2.6891 corresponds to σ in Megabarn $(10^{-18}\ cm^2)$ and the energy and the transition moment in atomic units. Alternatively, d^μ is a simplified form of (42a) of RLR in the case of FCSE approximation valid for open continuum. This establishes a link with the EAQDT formulation of RLR.

The power of FVVM is its simplicity. Each collision channel and each solution degenerate in energy corresponds to a term in expansion (3) and a solution of (4). The logarithmic derivative of the wave function at the boundary, instead of the R matrix appears naturally and explicitly in the equation (4). Moreover, with the present formulation, a link can be established between the system of eigen-equations in bound and in continuum spectrum. At a given continuum energy this logarithmic derivative is a discrete quantity which remplaces in the continuum the energy of the discrete spectrum.

3.- Application to the H₂ photoionization

The first application of the present method has been made on the photoionization of the one electron molecular system H_2^+ (see LRR). Here the method is applied to a prototype of many electron diatomic molecule, i.e. H_2 photoionization

$$H_2(X^1\Sigma_g^+) + h\nu \;\rightarrow\; H_2^+(X^2\Sigma_g^+) + e \begin{cases} \sigma_u \\ \pi_u \end{cases}$$

The internuclear distance is fixed at 1.4 a.u. and the calculation is performed in Frozen Core Static Exchange (FCSE) approximation. The expansion (2) contains in each channel six bound and two continuum (Coulomb) regular and irregular functions. Following Rudge (20), bound functions are of unique exponent but differ by the principal quantum number n. They are of Slater type with the exponent 2.0. Two asymptotic channels of different ℓ (ℓ =1,3) have been taken into account. In the table μ quantum defects and d^K transition moments (eq. 8a) obtained with FVVM are compared with the close coupling calculations using a standard method (21). Inspection of the Table reveals a good accurancy of quantum defects and transition moments within two significant figures of the exact close coupling results.

4.- Conclusion

The FVVM developed recently (LRR) and rederived in this paper has been applied with success to a two electron system photoionization showing in the case of this example no spurious resonances. The molecular photoionization cross section calculation is a stringent test as quantum defect and transition moments must be obtained. No such application has been performed starting from the R matrix theory. There exist a R matrix method for atomic photoionization due to Burke and Taylor (24) but the extension to molecules which should be straightforward has not yet been performed.

The comparaison between the present method and other Finite Volume Variational Methods can be performed using the Bloch (21) operator formalism as done by Lane and Robson (22) in nuclear physics. It can be shown that eigenchannel method (8-10) imposes more restrictive variational form of the wave function. The Jackson-Nesbet (11-12) method gives variational R matrix solving differently an equation similar to our equation (4b). The standard R matrix (1-7) obtains the collision quantities as representation of the R matrix on a set of functions. Only when Buttle and variational corrections (7) are added, the standard R matrix method gives results comparable with the other finite volume methods.

ACKNOWLEDGEMENTS

I am very grateful to Dr H. Le Rouzo for fruitful discussions and for making available to me his tricenter version of the Schaefer (25) program for integral calculation. I acknowledges Dr H. Lefebvre-Brion for critical reading of the manuscript.

| Electron Energy (a.u.) | μ | | | | d^K | | | |
| | σ_u | | π_u | | σ_u | | π_u | |
	FVVM	CC	FVVM	CC	FVVM	CC	FVVM	CC
0.00275	p 0.04474	0.04240	-0.092047	-0.9103	1.93750	1.94761	2.53543	2.51890
	f 0.00374	0.00430	0.002889	0.0031	0.01698	0.01842	0.01547	0.01522
0.0395	p 0.05242	0.05099	-0.086007	-0.0879	1.72858	1.75100	2.32759	2.31249
	f 0.00897	0.00470	0.002955	0.0034	0.02140	0.02287	0.02306	0.02673
0.1681	p 0.07882	0.07865	-0.077975	-0.07784	1.18824	1.22267	1.72466	1.73776
	f 0.00778	0.00530	0.003969	0.00392	0.03340	0.03512	0.03975	0.04074
0.3519	p 0.10454	0.10790	-0.064482	-0.06532	0.74305	0.76310	1.20529	1.20405
	f 0.00512	0.00617	0.004169	0.00460	0.03735	0.04064	0.03733	0.04753
0.5356	p 0.12392		-0.050704		0.47692		0.87425	
	f 0.00732		0.004078		0.03733		0.04621	
0.6459	p 0.12791		-0.049425		0.37165		0.74481	
	f 0.00833		0.006911		0.03976		0.04430	

REFERENCES

(1) E.P. Wigner and L. Eisenbud, Phys. Rev. 72, 29 (1947).
(2) P.G. Burke, A. Hibbert and W.D. Robb, J. Phys. B4, 153 (1971).
(3) B.I. Schneider, Phys. Rev. A11, 1957 (1975).
(4) P.G. Burke, I. Mackey and I. Shimamura, J. Phys. B10, 2497 (1977).
(5) T.K. Holley, S. Chung, C.C. Lin and E.T.P. Lee, Phys. Rev. A26, 1852 (1982).
(6) P.G. Burke, C.J. Noble and S. Salvani, J. Phys. B16, L113 (1983).
(7) D.J. Zvijac, E.J. Heller and J.C. Light, J. Phys. B8, 1016 (1975).
(8) U. Fano and C.M. Lee, Phys. Rev. Lett. 31, 1573 (1973) ;
    C.M. Lee, Phys. Rev. A10, 584 (1974).
(9) Ch. Jungen, Invited papers of the XII ICPEAC, Gatlinburg, Tennessee, 1981,
    edited by S. Datz (North Holland, Amsterdam, 1982), p. 455.
(10) G. Raseev and H. Le Rouzo, Phys. Rev. A27, 268 (1983).
(11) J.L. Jackson, Phys. Rev. 83, 301 (1951).
(12) R.K. Nesbet, Variational Methods in Electron Atom Scattering Theory, Plenum
    Press, 1980.
    R.K. Nesbet, J. Phys. B14, L415 (1981).
(13) C.H. Greene, Phys. Rev. A to be published.
(14) H. Le Rouzo and G. Raseev, Phys. Rev. A28 (1983) in press.
(15) W. Kohn, Phys. Rev. 74, 1763 (1948).
(16) C.B. Moler and G.W. Stewart, Siam J. Numer, Anal. 10, 241 (1973).
(17) J.H. Wilkinson, The Algebraic Eigenvalue Problem (Oxford University Press,
    London, 1965).
(18) M.R.H. Rudge, J. Phys. B6, 1788 (1973) ; 8, 940 (1975) ; 9, 2357 (1976) ;
    13, 3717 (1980).
(19) G. Raseev, Comput. Phys. Comm. 20, 267 (1980).
(20) P.G. Burke and K.T. Taylor, J. Phys. B8, 2620 (1975).
(21) C. Bloch, Nucl. Phys. 4, 503 (1957).
(22) A.M. Lane and D. Robson, Phys. Rev. 151, 774 (1966).
(23) H.F. Schaefer, J. Chem. Phys. 52, 6241 (1970) ;
    W.H. Miller, C.A. Slocomb and H.F. Shaefer, ibid 56, 1347 (1972).

A THEORETICAL STUDY OF THE H_2^- SYSTEM

AT LOW ENERGIES

George V. Nazaroff
Department of Chemistry
Indiana University at South Bend
South Bend, Indiana 46634 USA

An open-shell configuration interaction (CI) calculation was carried out for the $^2\Sigma_u^+$ and the $^2\Sigma_g^+$ states of the H_2^- system, and a number of adiabatic potential energy curves was obtained for internuclear separations between 1.05 and 4.00 au. The goal of this work was to search for possible H_2^- resonances within about a 10 eV energy region above the ground target state.

Computational details

The calculation used twenty configurations consisting of antisymmetrized products of three prolate spheroidal orbitals and a doublet spin function. The orbitals $1\sigma_g$, $1\sigma_g'$ and $1\sigma_u$ were used to describe the H_2 target, while the orbitals $n\sigma_g$ and $n\sigma_u$, $n = 2,...,6$, were used to describe the third electron. All of the orbitals were taken from the work of Eliezer, et al. (1).

For purposes of the following discussion it is convenient to arrange the orbitals according to increasing size. A measure of an orbital's size was obtained by arbitrarily setting the internuclear separation at 1.95 au, plotting the orbital along the direction of the internuclear axis (the z-axis), and locating the maximum value of the orbital. The results are given in Table I below, where the angular momentum symmetry type of each orbital in the united atom limit is also given.

TABLE I. Location of orbital maxima along
z-axis for internuclear separation of 1.95 au

orbital	z_{max} (au)	symmetry
$5\sigma_g$	2.2	s,d
$2\sigma_g$	2.3	s
$4\sigma_g$	6.3	s,d
$6\sigma_g$	7.1	s
$3\sigma_g$	8.8	s
$2\sigma_u$	on nuclei	p
$3\sigma_u$	5.8	p
$4\sigma_u$	11.3	p
$5\sigma_u$	24.6	p,f
$6\sigma_u$	43.1	p,f

The potential curves were calculated using the method of Harris (2), except that up to a thousand integration points were used in evaluating the electron-electron repulsion integrals with Simpson's rule. The curves for the $^2\Sigma_u^+$ and the $^2\Sigma_g^+$ systems are given in Figures 1 and 2, respectively. The potential curves of the H_2 parent target states, $X^1\Sigma_g^+$ and $b^3\Sigma_u^+$, are coincident with the lowest attractive and the lowest replusive H_2^- curves, respectively, on the scale of the graphs in these figures.

In Figures 1 and 2, as in all of the following figures, the plotted points represent the energies of the various CI roots. Smooth curves have been drawn somewhat arbitrarily by connecting those roots whose wave functions have similar dominant configurations.

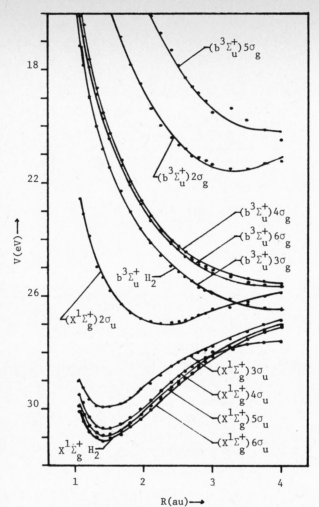

Figure 1. Adiabatic potential curves for the $^2\Sigma_u^+$ states of the H_2^- system

    The results in Figures 1 and 2 indicate that the CI roots line up in energy according to the size of the orbital in the dominant configuration containing the third electron. The shapes of the curves for which the dominant configurations contain either the $3\sigma_u$, $4\sigma_u$, $5\sigma_u$, $6\sigma_u$, or the $4\sigma_g$, $6\sigma_g$, $3\sigma_g$ orbital are similar to the shapes of their H_2 parent target states. These curves represent, no doubt, a free electron localized at a distance from the target.

    The results in Figures 1 and 2 also suggest that the first excited H_2 target state, $b^3\Sigma_u^+$, does not bind an electron in the Franck-Condon region of the ground H_2 target state. All H_2^- potential curves having the $b^3\Sigma_u^+$ H_2 target state as their parent lie above the parent state in this region of internuclear separations, in agreement with the findings of Buckley and Bottcher (3).

    The fourth potential curve in Figure 2, labeled as $(X^1\Sigma_g^+)2\sigma_g$, is more or less similar in shape to that of its parent target state. This root represents an s-wave electron localized near the H_2 target.

    The fifth root in Figure 2, labeled $(X^1\Sigma_g^+)5\sigma_g$, has the ground H_2 target state

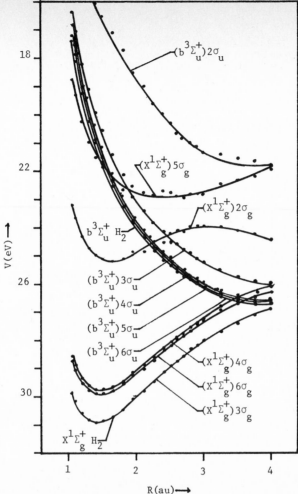

Figure 2. Adiabatic potential curves for the
$^2\Sigma_g^+$ states of the H_2^- system

as its parent, yet the shape of its potential curve is substantially different from that of its parent. Configurations involving the $5\sigma_g$, $4\sigma_g$ and $2\sigma_g$ orbitals are important for this root, thus suggesting that it represents a d-wave electron trapped in the field of ground H_2 target state.

Judging from the curves in Figures 1 and 2, the roots which could possibly represent low-lying H_2^- shape resonances are: $(X^1\Sigma_g^+)2\sigma_u$ in the $^2\Sigma_u^+$ system, which no doubt corresponds to the well-established p-wave shape resonance (4); and $(X^1\Sigma_g^+)5\sigma_g$ in the $^2\Sigma_g^+$ system which could possibly be a d-wave shape resonance. The root $(X^1\Sigma_g^+)2\sigma_g$ in Figure 2 is disqualified, since it would have to be an s-wave shape resonance, which does not seem to make physical sence.

Stabilization study

The two above mentioned roots were studied with the stabilization method (5) in order to evaluate their status as possible shape resonances. The stabilization method was implemented by cumulatively eliminating configurations from the CI matrix and thus progressively reducing the ordinal number of the root under investigation. The typical experience in all of the stabilization calculations was that the root under investigation sank in energy as configurations were deleted, but that all of the other remaining roots in the calculation hardly changed their energies at all.

The stabilization method was first applied to the $(X^1\Sigma_g^+)2\sigma_u$ root from Figure 1 in order to see how the method would treat a root already believed to represent a known shape resonance. The results are given in Figure 3 below. One can see, that after an initial drop in energy of 1 - 1.5 eV, this root stabilized, thus fulfilling the stabilization criterion for representing a resonance.

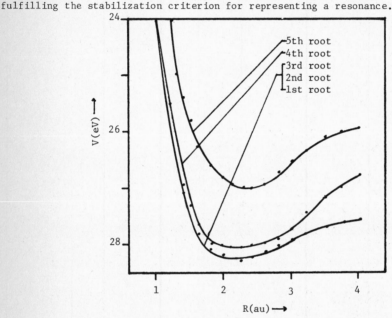

Figure 3. Stabilization study of the $(X^1\Sigma_g^+)2\sigma_u$ root.

The stabilization method was then applied to the $(X^1\Sigma_g^+)5\sigma_g$ root from Figure 2. The results are given in Figure 4 below. In contrast with the previous root, this one does not stabilize and therefore most likely does not represent an H_2^- resonance.

Note that in both the $(X^1\Sigma_g^+)2\sigma_u$ and the $(X^1\Sigma_g^+)5\sigma_g$ cases shown in Figures 3 and 4, respectively, omitting a small orbital from the CI calculation significantly affects the energy of the root under consideration, but omitting a large orbital does not affect the energy very much. This is a typical behavior encountered in stabilization studies (5).

Conclusion

The stabilization study, with the presently used basis set, verifies that the H_2^- system does have a resonance of $^2\Sigma_u^+$ symmetry formed by trapping a p-wave electron in the field of the ground H_2 target state. The study indicates that there is no comparable low-lying resonance having $^2\Sigma_g^+$ symmetry. The study also suggests that the first excited H_2 target state, the $b^3\Sigma_u^+$ state, does not bind an electron at moderate internuclear separations.

The results of the stabilization study can be conclusive only in the event that the orbitals used in the calculation actually do adequately span configuration space

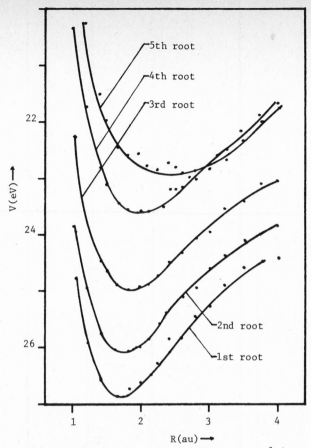

Figure 4. Stabilization study of the $(X^1\Sigma_g^+)5\sigma_g$

in the vicinity of the target molecule. Since the σ_u orbitals and the σ_g orbitals were themselves obtained as successive optimized roots of a variational calculation (1), presumably these functions do adequately span configuration space near the H_2 target. Presumably additional σ_u and additional σ_g orbitals would have their maximum densities at greater distances from the target than do the orbitals used in this study, and presumably these additional functions would not alter the stabilization results shown in Figures 3 and 4.

References

1. I. Eliezer, H. S. Taylor and J. K. Williams, Jr., J. Chem. Phys. 47, 2165 (1967)
2. F. E. Harris, J. Chem. Phys. 32, 3 (1960)
3. B. D. Buckley and C. Bottcher, J. Phys. B: Atom, Molec. Phys., 10, L635 (1977)
4. J. N. Bardsley, A. Herzenberg and F. Mandl, Proc. Phys. Soc., London 89, 305 (1966)
5. A. U. Hazi and H. S. Taylor, Phys. Rev. A1, 1109 (1970)

DISSOCIATIVE ATTACHMENT OF H$_2$ AND ITS ISOTOPE
BY LOW ENERGY ELECTRON IMPACT

Sujata Bhattacharyya
Gokhale College, Calcutta 700 020, India

Lali Chatterjee
Jadavpur University, Calcutta 700 032, India

1. Introduction

Dissociative attachment (DA) of H$_2$ by low energy electrons form an effective source of H$^-$ ions. Earlier work in this field upto 1973 has been reviewed by Schulz [1]. More recently this has been studied theoretically by Nesbet [2] and others.

We report here preliminary results of the electron impact DA of H$_2$, D$_2$ and (HD) in the frame work of quantum electrodynamics [3]. This technique has been successfully applied in some previous cases [4].

2. Theory

The reactions under consideration are

$$H_2 + e^- \rightarrow H^- + H \qquad (1)$$

$$D_2 + e^- \rightarrow D^- + D \qquad (2)$$

The target molecule is taken in $(1s\,\sigma_g)^2\ ^1\Sigma_g^+$ ground electronic state as this is stable and most abundant in nature. The DA time for the molecule is known to be much shorter than the rotational and vibrational times. Hence it is assumed that the internuclear axis of the molecule remains fixed during the entire collision process. In particular we take the molecule in ground rotational and vibrational state. In this state hydrogen molecule at equilibrium consists almost exclusively of para form. We use Coulomb gauge, and bound state vectors are written in a field theoretic way [4].

In Coulomb gauge and for low energy encounters the dominant term of the S-matrix comes from the static Coulomb interaction.

$$H_C = \frac{\rho(x)\rho(x')}{x - x'}\ d^3x\ d^3x' - \frac{\rho(x)\sigma(x')}{x - x'}\ d^3x\ d^3x' \qquad (3)$$

where $\rho(x)$ and $\sigma(x)$ are the charge densities for electron and proton field respectively. The cross-section for the DA can be written as

$$\sigma_{DA} = \frac{1}{(2\pi)^6}\ |M_{fi}|^2\ \frac{d^3 Q_A\ d^3 Q_{H^-}}{|P_e|} \qquad (4)$$

The projectile electron has momentum P_e while Q_A and Q_{H^-} are the momenta of the reaction products, i.e. atom and ion respectively. The initial state vector for the interacting particles consisting of electron and molecule is written as [4]

$$|\psi_i\rangle = \sum_{a\,b\,c\,x\,y} \int g_{a\,b\,c\,x\,y}^{\,d\,p\,r\,q\,s}(\ell_1\,\ell_2\,\ell_3\,P_4\,P_5)\, a_\alpha^\dagger(\ell_1)\, a_\beta^\dagger(\ell_2)\, a_\gamma^\dagger(\ell_3)\, B^\dagger(P_4)\, B^\dagger(P_5)|0\rangle \qquad (5)$$

$$|\psi_f\rangle = \sum_{\substack{def x' y'}} \int g_{\substack{def x' y'}}^{P\sigma\lambda x_i s'} (\ell_i' \ell_2' \ell_3' q_4 q_5) \, a_\rho^\dagger(\ell_i') \, a_\sigma^\dagger(\ell_2') a_\lambda^\dagger(\ell_3') \, B_x^\dagger(q_4) \, B_s^\dagger(q_5)|o\rangle \tag{6}$$

The g-fn. is the Fourier transform in momentum space of the solution of Schrodinger equation for the system of particles contained either before or after the dissociative attachment as the case may be. a^s and B^s are the annihilation operators for electrons and protons with ℓ, ℓ' and P, q as their respective momentum variables.

After making necessary substitution for g-fn. and neglecting Coulomb distortion part of the projectile electron because of low energy encounters and integrating out over the momentum variables.

$$M_{fi} = \frac{1}{2\pi} \langle \psi_f | H_e | \psi_i \rangle = M_{ee} - M_{ep} \quad . \tag{7}$$

and

$$M_{ee} = NS \sum_{\lambda=1}^{3} A_\lambda I_{\lambda g} \quad M_{ep} = NS' \sum_{\lambda=1}^{3} A_\lambda' I_\lambda' \tag{8}$$

where

$$N = \frac{3(2\pi)^3}{r_e^2} \left[\exp(r_e/2) - \exp(-r_e/2) \right] (2 + 2S)^{-\frac{1}{2}}$$

$$S = (1 + r_e + 1/3 \; r_e^2) \exp(-r)$$

r_e = equilibrium separation of the nuclei

A and A' are the factors arising out of the product of the electron spins and S, S' are the similar factors for nuclear spins.

$$I_1 = \delta^3(P_e - Q_c) \frac{16}{q\sqrt{\pi} (1 + P_e^2)^2} \int \psi_N(r) \exp(i \; r \; s) d^3 r \tag{9}$$

$$I_2 = I_3 = \delta^3(P_e - Q_c) \frac{24}{\sqrt{\pi}} (1 + P_e^2)^{-1} (q + P_e)^{-2} \int \psi_N(r) \exp(irs) d^3 r$$

$\psi_N(r)$ = nuclear w. function in ground rotational and vibrational state.

The integrals are performed using the properties of the error function with complex arguments. Similarly for

$$I_1' = \delta^3(P_e - Q_c) \frac{64}{9\sqrt{\pi}} (1 + P_e^2)^{-2} \left[9 + (2S + P_e)^2 \right]^{-1} \int \psi_N(r) \, Exp\left\{ -\frac{r}{2} - i(S + \frac{P_e}{2})r \right\} d^3 r$$

$$I_2' = \delta^3(P_e - Q_c) \frac{64}{.9\sqrt{\pi}} (1 + P_e^2)^{-1} \int \psi_N(r) \, Exp\left\{ -\frac{r}{2} + i(2S + \frac{P_e}{2})r \right\} d^3 r$$

$$\tag{10}$$

$$s = Q_A - Q_H$$

After averaging over the initial spins and summing over the final spin states of the interacting systems, we get some numerical factors and the probability amplitude becomes.

$$| M_{fi} |^2 = 3 N^2 \left[\sum_{\lambda=1}^{3} I_\lambda + \sum_{i=1}^{3} I_i' \right] \tag{11}$$

3. Discussion

DA is computed for the formation of H^- and D^- for projectile

energy between the threshold of the reaction channel 3.73 eV to
16 eV. The curve (Fig.1) obtained by normalising our results with
existing experiments [5] show resonance character near 12 eV. Ener-
getically this is the range where valence excited resonance $(1s\sigma_g)$
$(2P\sigma_u)^2\,{}^1\sum_g{}^+$ of molecular ion may occur. However unlike experimental curve we find monotonic decrease of DA cross-section beyond 13 eV. The reason behind may be absence of Rydberg excited
states of the reaction products in our case. As we are interested
in the ground states of H-atom and its negative ion, our result is
in fair agreement with experiment and other theories. We have also
computed (Fig.2) for H^- formation from HD. Though σ_{DA}^{HD} shows a
peak for D^- near the threshold of dissociative attachment, it
corresponds to shape resonance of $(1s\sum_g)^2(2P\,\sigma_u)^2\sum_u{}^+$. It
is worth comparing our result with that of experiment by Schulz
and Asundi [6].

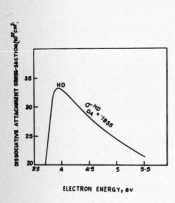

Fig. 1

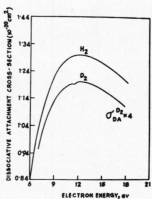

Fig. 2

FIGURE CAPTION

Fig. 1. Present work on dissociative attachment cross-sections of
H_2 and D_2 for the formation of H^- and D^- ions in the
ground state.

Fig. 2. Present work on dissociative attachment cross-section
of HD for the formation of H^- in the ground state.

ACKNOWLEDGEMENTS

L. Chatterjee thanks CSIR for financial help. Authors
thank T. Roy for his interest in the work.

REFERENCES

1) C.J. Schulz, Rev. Mod. Phys. 45 (1973) 423.

2) R.K. Nesbet, ICPEAC Abstract 1 (1981) 419.

3) T. Roy, Lett. Al Nuovo Cim. 5 (1972) 1045.

4) Sujata Bhattacharyya, T. Roy, Calcutta Math. Soc., 70 (1980) 89.
   S. Bhattacharyya, L. Chatterjee, T. Roy, Physica 106C (1981) 135.
   L. Chatterjee, S. Bhattacharyya, Phys. Letts. A93 (1983) 360.
   T. Roy and S. Bhattacharyya, Lett. Al Nuovo Cim. 9 (1974) 54.

5) D. Rapp., T.E. Sharp, D.D. Briglia, Phys. Rev. Lett. 14 (1965) 533.

6) G.J. Schulz and R.K. Asundi, Phys. Rev. 158 (1967) 25.

NUMERICAL ASYMPTOTIC FUNCTIONS IN CALCULATIONS
OF ELECTRON-MOLECULE SCATTERING

R. K. Nesbet

IBM Research Laboratory, San Jose, California 95193, USA

Model studies show that variational R-matrix theory and hybrid versions of the Kohn variational method can be organized so that principal computational steps are common to both methods. In order to take advantage of these theoretical developments, a collaboration has been initiated that includes P. G. Burke, at Belfast; C. J. Noble, at the Daresbury Laboratory, England; L. A. Morgan, at Royal Holloway College, England; and the present author. The general plan is to explore and design computational methods for electron-molecule scattering that can be built onto the existing ALCHEMY system[1] of bound state molecular wave function programs. The programs are at present limited to diatomic target molecules, using atomic basis orbitals of exponential form (STOs), augmented by continuum basis functions in the form of spherical Bessel functions and by numerical asymptotic functions (NAFs). Integrals required for calculations on polyatomic target molecules are currently being programmed.

Methods and programs are being tested in calculations of $e^- + H_2$ scattering. Variational R-matrix calculations were carried out within the boundary radius $r_1 = 10.0$ a_0, then matched to external solutions of the asymptotic coupled differential equations. The target H_2 ground state, $^1\Sigma_g^+$, is represented by the SCF wave function of Fraga and Ransil,[2] with a total of six σ atomic basis orbitals. An additional six π STOs were added in calculations taking $\sigma \rightarrow \pi$ polarizability into account.

In the static exchange approximation, only the target ground state is considered, represented by its static quadrupole potential in the asymptotic region. Converged results were obtained for coupled s and d partial waves in the $^2\Sigma_g^+$ scattering state, using only eight continuum basis functions in the form of NAFs and Bessel functions for k^2 up to $0.36a_0^{-2}$. Earlier calculations, using only STOs on the scattering center with no true continuum basis, had required 13 such basis functions for convergence. Results for K-matrix elements, computed for the $^2\Sigma_g^+$ scattering state, are listed in Table 1. Earlier results of Collins et al.[3] are given for comparison. Compared with typical accurate bound state calculations, the orbital basis set used here is small, giving only 15 configuration state functions for the variational R-matrix basis. All calculations here are for internuclear separation 1.402 a_0.

Experience with electron-atom scattering calculations indicates that the polarization response of the target system to an incoming electron can be taken into account by including polarization pseudostates in the close-coupling formalism. The SCF ground state wave function of H_2 is the single configuration $(1\sigma_g^2)\,{}^1\Sigma_g^+$. Pseudostate orbitals $\bar{\pi}_u$ and $\bar{\sigma}_u$ were computed variationally and were used to construct two pseudostate wave functions

$$(1\sigma_g\bar{\pi}_u)^1\Pi_u \, , \quad (1\sigma_g\bar{\sigma}_u)^1\Sigma_u^+ \, . \tag{1}$$

For calculations involving the $^1\Pi_u$ pseudostate or manifold of virtual excitations, six basis STOs of π symmetry were added to the Fraga-Ransil σ orbital basis. The computed parallel and transverse polarizabilities are respectively

$$\alpha_{\parallel} = 7.0149 \; a_0^3 \, , \quad \alpha_{\perp} = 4.9660 \; a_0^3 \, , \tag{2}$$

to be compared with accurate values (at R=1.40 a_0),[4]

$$\alpha_{\parallel} = 6.3805 \; a_0^3 \, , \quad \alpha_{\perp} = 4.5777 \; a_0^3 \, . \tag{3}$$

In order to include all components of the dipole transition moment from the $^1\Sigma_g^+$ ground state, the following scattering channels must be represented, if p and f partial waves are included in the ground state channels:

Target State	Partial Wave Orbital
$^1\Sigma_g^+$	$p\sigma_u \quad f\sigma_u$
$^1\Pi_u$	$d\pi_g \quad g\pi_g$
$^1\Sigma_u^+$	$s\sigma_g \quad d\sigma_g \quad g\sigma_g$

$$\tag{4}$$

Thus, a 2-channel static exchange calculation must be extended to seven channels for a consistent and comparable polarized pseudostate calculation. Calculations of this kind were carried out at R=1.402 a_0. Results for open-channel K-matrix elements are summarized in Table 2, and are compared with earlier calculations by Klonover and Kaldor,[5] who represented dynamical polarization effects by a second-order optical potential. In the present calculations, all components of dipole transition moments and of quadrupole moments were computed for the three target states. This matrix of moments was used to construct the asymptotic potential functions for integration of coupled equations outside the R-matrix boundary.

The continuum orbital basis in these calculations was carried to effective completeness by successively adding Bessel functions in each of the partial wave channels.

Alternative ways of including numerical asymptotic functions in the orbital basis set were tested. While two independent NAFs per open channel must be used to represent oscillatory functions over an extended range of the radial variable, the values of k and ℓ in the present calculations are such that the open-channel solutions of the asymptotic equations have few if any oscillations inside r_1, the R-matrix boundary. In order to avoid linear dependency problems one regular vector NAF per open channel was constructed as a linear combination of the numerically integrated asymptotic functions. The diagonal component of this function was fitted to a regular spherical Bessel function at a matching radius r_0, 2.0 a_0 in the present case, and was used as a continuum basis function. Completeness of the continuum basis inside r_1 is achieved by adding Bessel functions that vanish at r_1, with successively greater numbers of radial modes. Bessel functions were added in each of the seven partial wave channels until effective convergence to two or three significant decimals was achieved in the eigenphase sum. Convergence was smooth and rapid, requiring only 42 configuration state functions for the results listed in Table 2. Wave functions of the same structure were used over the full energy range studied.

The computed eigenphase sums shown in Table 2 are in good agreement with the earlier work of Klonover and Kaldor,[5] although the individual K-matrix elements differ more evidently. The systematic sign reversal of element K_{31} is an artifact of a phase convention. The present calculations confirm the general trend of the Klonover-Kaldor eigenphase sums. This strengthens the argument[6] that the $e^- + H_2$ scattering resonance is accompanied by a strongly energy-dependent background, not considered in the current theory of dissociative attachment.

These are the first electron-molecule scattering calculations to include a full dynamical pseudostate model of molecular polarization. Special methods were developed to solve many technical problems associated with the asymptotic integration and construction of useable basis NAFs. These developments will be discussed in separate publications.

This work, at IBM San Jose, was supported in part by the U.S. Office of Naval Research.

REFERENCES

1. A. D. McLean, M. Yoshimine, B. Liu, and P. S. Bagus, ALCHEMY program package, unpublished, IBM Research Laboratory, San Jose, California.
2. S. Fraga and B. J. Ransil, *J. Chem. Phys.* **35**, 1967 (1961).
3. L. A. Collins, W. D. Robb, and M. A. Morrison, *Phys. Rev.* **A21**, 488 (1980).
4. W. Kolos and L. Wolniewicz, *J. Chem. Phys.* **46**, 25 (1967).
5. A. Klonover and U. Kaldor, *J. Phys.* **B12**, 3797 (1979).
6. R. K. Nesbet, *Comments on Atomic and Molecular Physics* **11**, 25 (1967).

TABLE 1

$e^- + H_2$ Static Exchange, K-Matrix Elements in the $^2\Sigma_g^+$ Scattering State

$k(a_0^{-1})$	K_{00} a	b	K_{20} a	b	K_{22} a	b
0.1	−0.2172	−0.217	0.3589(−2)	0.39(−2)	0.2248(−2)	0.21(−2)
0.2	−0.4504	−0.451	0.7107(−2)	0.73(−2)	0.4346(−2)	0.45(−2)
0.3	−0.7227	−0.722	0.1038(−1)	0.11(−1)	0.7237(−2)	0.74(−2)
0.4	−1.0661	−1.07	0.1316(−1)	0.13(−1)	0.1107(−1)	0.11(−11)

[a]Present results, R=1.402 a_0.

[b]Collins *et al.*, Reference 3.

[c]$0.n \times 10^{-m}$ is written as $0.n(-m)$.

TABLE 2

$e^- + H_2$ Polarized Pseudostate, K-Matrix Elements and Eigenphase Sums in the $^2\Sigma_u^+$ Scattering State

$k(a_0^{-1})$	K_{11} a	b	K_{31} a	b	K_{33} a	b	$\Sigma\eta$ a	b
0.1	0.2493−1		0.1123−2		0.1662−2		0.2659−1	
0.2	0.9422−1		0.2391−2		0.4415−2		0.9836−1	
0.3	0.2356	0.1827	0.3973−2	−0.1165−2	0.7688−2	0.1629−2	0.2391	0.1823
0.4	0.4549	0.3885	0.6350−2	−0.7644−2	0.1111−1	0.6913−2	0.4380	0.3774
0.5	0.7211	0.6225	0.9697−2	−0.2828−2	0.1632−1	0.1710−1	0.6410	0.5739
0.6	0.9939	0.9455	0.1357−1	−0.1542−1	0.2451−1	0.3056−1	0.8068	0.7878
0.7	0.1238+1	0.1220+1	0.1696−1	−0.3831−1	0.3261−1	0.4393−1	0.9236	0.9272
0.8	0.1400+1	0.1489+1	0.2263−1	−0.3023−1	0.3775−1	0.5358−1	0.9882	0.1033+1
0.9	0.1519+1	0.1673+1	0.2776−1	−0.1572−1	0.4304−1	0.6026−1	0.1031+1	0.1092+1
1.0	0.1591+1		0.3249−1		0.5196−1		0.1061+1	

[a]Present results, R=1.402 a_0

[b]Data from Klonover and Kaldor, Reference 5, R=1.40 a_0.

PART III

COUPLING NUCLEAR MOTIONS
WITH THE IMPINGING ELECTRON:
THE THEORETICAL MODELS

RESONANT AND NON-RESONANT COUPLINGS TO THE NUCLEI IN

ELECTRON-MOLECULE COLLISIONS.

A.Herzenberg
Applied Physics, Yale University,
New Haven, Connecticut 06520, USA.

ABSTRACT.
    This is an introduction to the papers by Gauyacq, Domcke,
and Temkin. The concepts of 'resonance' and 'virtual state' are
defined. The concept of a 'virtual state' in the presence of a
weak dipole-moment is discussed briefly.

We shall confine ourselves to electron-molecule interactions dominated
by short distances, no more than 3 or 4 bohr-radii. The reactions then
depend sensitively on the amplitude of the wavefunction of a bombarding
(or departing) electron inside a target molecule, or very close to it. This
is quite different from collisions at longer range where the interaction
between an electron and its target is due to the long range coulomb
interaction. For our purposes, the collision energies will be mostly below
ten electron-volts.

An important benchmark is Massey's (1935) calculation of the cross-
section for vibrational excitation in the collision of an electron with a
molecule without a dipole moment. He obtained a value of the order of 10^{-18}
cm^2 for the total cross-section, integrated over all scattering angles.
The calculation was done in the Born approximation, that is by neglecting all
distortion of the wavefunction of the projectile by the interaction with the
target. Because there were no dipole moments, there was no long range
interaction to affect the electron outside the molecule. The observed
vibrational excitation cross-sections turn out to be larger than Massey's
benchmark by up to three orders of magnitude, within limited regions of
energies, of the order of 1 ev wide at resonances and at thresholds.

There are two types of mechanism - resonances and virtual states - which
can enhance the amplitude of an electron's wavefunction at a molecule by the
required factor 10 or more over the plane wave used by the Born approximation.
For our present purposes, a resonance is almost always a state localised
within a centrifugal potential barrier at a positive collision energy; the
amplitude enhancement comes from a constructive interference of many waves due
to reflection at the inside of the potential barrier, after the wave-function
has leaked in from the outside. Figure 1 illustrates a resonance in a
potential well within a barrier; the diagram is taken from Fermi's lecture
notes on nuclear physics (Fermi, 1948).

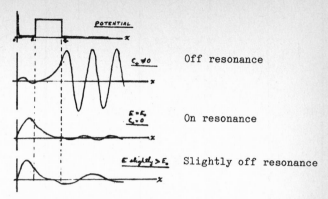

Off resonance

On resonance

Slightly off resonance

Figure caption.

The diagram shows amplitude enhancement by a resonance in a one-dimensional potential well with a barrier. The enhancement occurs only over a small range of energies where the multiply reflected waves within the well interfere constructively. (Fermi, 1948).

The prototype of a resonance in electron-molecule scattering is e + N2 at an incident energy of 2 ev, which has been discussed in an enormous number of papers. For the e + N2 resonance, several different theoretical approaches agree more or less with one another, and with a very large amount of experimental data on the excitation of up to 10 quanta of vibrations from initial states which may also be vibrationally excited. The resonance is a quasistationary state localised by the centrifugal potential due to a d-wave outside the molecule. (For recent references, see Burke(1979), Lane(1980), Hazi et. al.(1981), Kazansky(1978-1983)).

In an electron-molecule resonance, the energy and width depend on the positions of the nuclei. If one treats the complex resonance energy as a function of the nuclear positions, and uses it as a potential acting on the nuclei during the life of a temporary negative ion, one gets the 'local complex potential model'. Domcke's paper (in this volume) discusses improvements from going beyond that model, to a non-local model, which effectively takes the finite velocities of the nuclei into account. Temkin's paper (also in this volume) describes another approach for avoiding the local approximation, by a close-coupling expansion in vibrational nuclear wavefunctions; although this method tends to be heavy on computer time, it may be particularly useful near thresholds, where the local model is suspect.

Enhancement by a 'virtual state' occurs if the s-wave of the projectile constitutes a major component of the wavefunction, and if a new bound state with a large s-wave component is on the verge of being born; there may then be a large amplitude enhancement from the fact that the wavefunction outside the target at low energies is $\cos(kr)/(kr)$, instead of $\sin(kr)/(kr)$ as in the incident wave; this is an enhancement of order $1/(k \times$ target radius) for very low k. Since the s-wave predominates, there is no dominant centrifugal potential barrier as there is at resonances; the localisation represented by the factor $(1/r)$ is very weak compared with the strong localisation due to the centrifugal potentials at resonances. Such 'amplitude enhancement without trapping' appears to occur frequently near thresholds.

The prototype of the virtual state mechanism is the excitation of the vibrational breathing mode in the reaction e + SF6 at incident energies below 1 ev; the observed cross-sections show strong threshold spikes extending up to about 1/2 ev above threshold. The literature on this problem at the vibrational thresholds is almost entirely experimental; the theory has not been well developed because the limelight has been hogged by molecules such as HCl, with weak dipole moments about whose effects there is no general agreement. SF6 is important in part because its symmetry forbids a dipole moment, so that it represents a case of strong vibrational excitation at threshold by an s-wave electron which is not distorted by the potential field of a dipole moment.

At a threshold showing a virtual-state enhancement, the shape of the wavefunction of the projectile electron within the molecule changes little over an energy of the order of an electron-volt; therefore the logarithmic radial derivative may be taken to depend only on the positions of the nuclei, and to be independent of the energy. Gauyacq's paper (in this volume) discusses applications of this approximation, which avoids a detailed calculation inside the molecule.

The effect of a dipole-tail in the electron-molecule potential on the virtual-state enhancement has recently been the subject of much speculation. The virtual-state enhancement is connected with a pole in the scattering amplitude, which in its turn is associated with the new bound state on the verge of being born. Domcke and Cederbaum (1981) showed that in the presence of a dipole tail, the track of this pole in the complex momentum plane follows the positive imaginary axis down to the origin, and stops there; unlike the track for a short range potential, it does not continue down the negative imaginary axis. This effect is important because it is the portion of the track in the lower half plane which is responsible for the virtual state enhancement with a short-range potential when there is no bound state.

Saha and I have done a calculation which sheds more light on the problem of the apparent vanishing of the pole from the lower half plane (Herzenberg and Saha, 1983). The dipole tail gives the scattering amplitude a branch point at the origin in the complex momentum plane; therefore several sheets of the Riemann surface have to be studied. Domcke and Cederbaum confined their calculation to the 'physical sheet', which they define by a cut down the negative imaginary axis. Saha and I went beyond their calculation by looking for the pole also on the adjacent Riemann sheets, which are still close enough to the physical momentum axis to affect the observed scattering amplitude. We find that the pole trajectory does continue from the upper into the lower half of the complex momentum plane, if the dipole moment is not too large; in fact the trajectory splits at the origin into two trajectories which continue into the lower half plane on the two Riemann sheets on either side of the sheet studied by DC. The amplitude enhancement on the physical momentum axis at the birth of a new bound state remains as strong as in the absence of a dipole moment.

References:

Burke P.G.,
1979, Advances in Atomic and Molecular Physics, 15, 471.

Domcke W., and Cederbaum L.S.,
1981, J.Phys. B. (Atom. and Mol. Physics) 14, 149.

Domcke W.,
1981, J.Phys. B. (Atom. and Mol. Physics) 14, 4889.

Fermi E.,
1948, Lectures on Nuclear Physics, University of Chicago.

Hazi A.U., Rescigno T.N., and Kurilla M.,
1981, Phys. Rev. 23, 1089.

Herzenberg A., and Saha B.C.,
1983, J.Phys. B. (Atom. and Mol. Phys.) 16, 591.

Kazansky A.K.,
1978, Teor. Mat. Fyz., 36, 414; 37, 84;
1980, Some Aspects of Atomic Collision Theory
        (Leningrad: Leningrad State University);
1982, Zh. Eksper. Teor. Fyz. 82, 1422;
1983, J.Phys. B. (Atom. and Mol. Phys.) 16, 2427.

Lane N.F.,
1980, Rev. Mod. Phys., 52, 29.

Massey H.S.W.,
1935, Trans. Faraday Soc., 31, 556, 1935.

NON RESONANT TREATMENT OF ASSOCIATIVE DETACHMENT AND DISSOCIATIVE ATTACHMENT

J.P. Gauyacq

Laboratoire des Collisions Atomiques et Moléculaires[†], Université Paris-Sud
Bât. 351, 91405 ORSAY Cedex, FRANCE

The dissociative attachment (DA) and associative detachment (AD) processes are associated with energy exchange between the electronic and nuclear motions. In both processes, one of the particle (e^- or nuclei) makes a transition from a discrete to a continuum state, the other particle undergoing the reverse transition :

$$A^- + B \leftrightarrow AB(v,J) + e^-$$

Until recently, these processes were described in a resonant formalism i.e. by invoking the formation during the collision time of an unstable negative ion AB^- which decays either by dissociation or by electron emission. Presently, a non resonant formalism is presented, which does not *explicitly* consider a resonant AB^- state and does not make any local approximation of its decay.

The low energy e^--static molecule interaction is described by an effective range theory (ERT) : far away from the molecule ($r > r_c$), the electron feels a long range potential, and the short range forces are represented by a boundary condition at $r = r_c$, that is taken independant of the electron energy in the low energy limit. The boundary condition can be either fitted to reproduce experimental data or extracted from *ab initio* calculations. The ERT formalism extends the zero range potential approximation (ZRP)[1,2] and is analogous to the quantum defect theory. The interaction between the electronic and vibrational motions is thus confined to the boundary condition. Similarly to the ZRP approximation[2], the electron-vibration interaction in the course of the collision can be treated exactly[3] : the total wavefunction is a superposition of products of unperturbed target states by free electron scattering states, and the coefficients of this superposition are the solutions of a set of linear algebraic equations. For the AD process, the solution of a small set yields the ratio between the amplitudes of the open channels ($AB(v,J) + e^-$) i.e. of the cross sections. A large set of equations provides the description of the $A^- + B$ channel as a superposition of a large number of ($AB(v,J) + e^-$) open and closed channels, and therefore the normalization and the absolute value of the cross sections[2] (in other words, the DA process is described as vibrational excitation to closed channels). In addition, for low energy of the relative motion in AD, the detachment probability P_d is close to unity as soon as the system can enter into the unstable region ; the results presented below have been obtained with the approximation of the detachment probability P_d set to one, inside the Langevin radius. All the calculations were performed for the AD process, and the DA cross sections obtained through detailed balance.

Two applications are presented, corresponding to two different physical situations : the F^--H collision which is a predominantly s wave problem, and the e^--H_2 dissociative attachment at low energy which is associated with a p wave (see also the application to DA in e^--HCl collisions by D. Teillet Billy and J.P. Gauyacq at the same conference)[7].

- Associative detachment in the F^--H collisions : $F^- + H \rightarrow HF(v,J) + e^-$

In a pure s wave problem, the bound state of the $(AB)^-$ ion, when disappearing, transform into a virtual state, and in this case, it is not possible to define a resonant AB^- state (in the presence of a long range dipolar field, like in HF, the situation is modified[4]). The F^--H collisional system is studied with two different formalisms : first, the ZRP approximation which only considers short range forces and an ERT formalism with the long range dipolar field taken into account ; in both cases, the boundary condition has been extracted from the *ab initio* calculation of Segal and Wolf[5]. The ERT results for a 25 meV head on F^--H collision are presented on figure 1 : it displays the relative population of the final vibrational HF states.

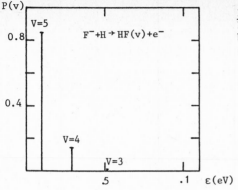

Fig. 1 : Relative population of the final $\overline{HF(v)}$ states in F⁻-H collision (25 meV, b=0). The abscissa scale is the emitted electron energy scale.

The main feature is the dominance of the highest v level, i.e. of the low energies of the emitted electron. This is a general feature of the s wave detachment problem at low collision energy ; the AB⁻ bound state disappears when touching the continuum and is not connected to any resonant state in the unstable region, and thus the bound electron is ejected towards the low lying continuum states. The two sets of results for the relative reaction rates at 300°K are presented on figure 2 together with experimental data[6]. They are very similar and in good agreement with the experimental results ; both shows a dominance of the high v levels. The similarity of the two sets of results shows that this v dependence is not due to the singularities of the dipolar field[3].

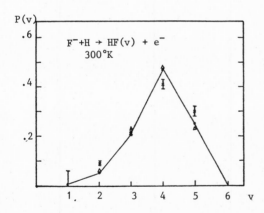

Fig. 2 : Relative population of the $\overline{HF\ (v)}$ levels formed in F⁻-H collision at 300°K - x : experiment[6] ; Δ : ZRP results , –o– ERT results.

– Dissociative attachment in low energy e⁻-H₂ collisions :

The e⁻-H₂ interaction is described in an ERT formalism : in the outer region, the electron feels the polarisation potential and the centrifugal potential. The boundary condition is expanded to second order around the equilibrium distance, thus needing 3 parameters : two of them were extracted from *ab initio* calculations[8] and the third one was adjusted so that the theoretical threshold value of attachment on

H_2 agrees with the experimental value of 1.6×10^{-21} cm^2.[9] The process was studied from the AD side and DA process was obtained through detailed balance. AD results for a 25 meV H$^-$-H collision are presented on figure 3 ; it displays the relative population of the final vibrational states in the process H$^-$+H $\rightarrow$ H$_2$(v) + e$^-$.

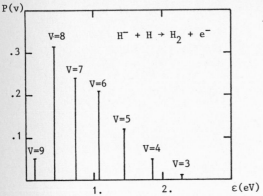

Fig. 3 : Relative population of the final H$_2$(v) levels in H$^-$-H (25 meV) head on collision. The abscissa scale is the emitted electron energy scale.

In contrast with the previous case, the population is spread over quite a few vibrational levels. Indeed, the system belongs to the p wave case and the H$_2^-$ bound state transforms into a resonance state at small R ; in contrast with the F$^-$-H system, the H$^-$-H system can penetrate into the continuum before emitting the electron and thus can emit higher energy electrons. Results on DA cross sections at threshold are presented on figure 4, it displays the ratio of the cross sections (25 meV above threshold) for various v levels, and for the H$_2$ and D$_2$ targets. Good agreement is found with the experimental results of Allan and Wong [10] and the theoretical results of Bardsley and Wadehra [11]. The isotope effect for ground state molecules at threshold is found to be $\sim$ 230, as compared with the experimental value of $\sim$ 200 of Schulz and Asundi [9]

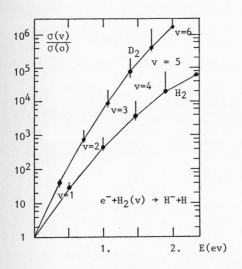

Fig. 4 : Ratio between the threshold attachment cross sections for various v levels. The abscissa scale is the internal energy scale of the initial target state.

As a conclusion, the ERT formalism provides a description of the low energy e^--molecule interaction, that can be linked with *ab initio* calculation and that is compact enough to allow for an exact treatment of interaction between the electron and the vibrational motion. Such formalisms are required for situations where the local description of the AD and DA processes in terms of resonant states fails, e.g. s wave problems, or extremely broad resonances.

†(laboratoire associé au CNRS)

Rererences

1. Demkov Yu N. 1964, Sov. Phys. JETP 19, 762 ; Dubé and Herzenberg 1977, Phys.Rev. Let. 38, 820 ; Fabrikant 1978, J.Phys.B 11, 3821 and 1983, 16, 1253.

2. Gauyacq J.P. 1982, J.Phys.B 15, 2721

3. Gauyacq J.P. 1983, J.Phys.B in press

4. Domcke W. 1981, J.Phys.B 14, 4889 ; Herzenberg and Saha 1983, J.Phys.B 16, 591 ; Estrada H. and Domcke W. 1983, to appear in J.Phys.B

5. Segal G.A. and Wolf K. 1981, J.Phys.B 14, 2291

6. Zwier T.S., Weisshaar J.C. and Leone S.R. 1981, J.Chem.Phys. 75, 4885 ; Smith M.A. and Leone S.R. 1983, J.Chem.Phys. 78, 1325

7. Teillet-Billy D. and Gauyacq J.P., Lecture notes in Chemistry, same volume

8. Gibson T.L. and Morrison M.A. 1981, J.Phys.B 14, 727 ; C.W. McCurdy and R.C. Mowrey 1982, Phys.Rev. A 25, 2529

9. Schulz G.J. and Asundi R.K. 1966, Phys. Rev. 188, 280

10. Allan M. and Wong S.F. 1978, Phys. Rev. Let. 41, 1791

11. Bardsley J.N. and Wadehra J.M. 1979, Phys. Rev. 20, 1398

ELECTRON DISSOCIATIVE ATTACHMENT ON HCl MOLECULES

D. Teillet-Billy and J.P. Gauyacq

Laboratoire des Collisions Atomiques et Moléculaires, Université Paris-Sud,
Bât. 351, 91405 ORSAY Cedex, FRANCE

The dissociative attachment (DA) of slow electrons ($E \lesssim 1eV$) on HCl molecules :

$$e^- + HCl \rightarrow Cl^- + H$$

has been shown experimentally by Allan and Wong [1] to be rapidly varying with the internal energy of the target molecule (vibrational or rotational energy). This process is associated with a $(Cl^- - H)$ state of $^2\Sigma^+$ symmetry at large distance, and the attached electron has predominantly a s wave symmetry. As a consequence, the HCl^- $^2\Sigma^+$ state, that is bound at large distances, cannot be correlated with a standard resonant state at small R. The usual picture of the DA process is no more applicable, one rather has to consider that the incoming electron induces some vibrational excitation in the target molecule, and when the nuclei are separated enough the electron gets captured into a bound state. We present a theoretical study of DA, that is able to describe such a process. It complements a previous study on the reverse reaction in associative detachment [3] , and also extends a previous study of dissociative attachment in the ZRP approximation which solved the collision problem by a first order perturbation [7] . The electron-static molecule interaction is described in an effective range theory (ERT), that is an extension of the zero range potential approximation [2,3,4] : when the electron is far from the molecule ($r > r_c$), it feels the polarisation potential of the molecule ; and the short range forces are taken into account via a boundary condition ar $r = r_c$ on the logarithmic derivative of the electron wavefunction. In this formalism, the electronic and vibrational motions are only coupled via the boundary condition. The HCl long range dipolar field has been neglected[4].

The collision equations are solved for the associative detachment process and the DA cross sections are obtained through detailed balance. The collision equations can be solved exactly in the same way as in the ZRP approximation[3], the total wavefunction is expressed as a close coupling expansion (total angular momentum J)

$$\Psi_J = \sum_N A_N^J \cdot \chi_N^J (R) \; \phi_{e\ell} (k_N^J, r) \; Rot(J)$$

where χ_N^J are the HCl vibrational functions (anharmonic)

$\phi_{e\ell}$ are the electronic wavefunctions, solutions of the outer region ($r > r_c$) with asymptotic outgoing wave behaviour.

The A_N amplitudes are solutions of the algebraic equations obtained by bringing the above expansion into the boundary condition at $r = r_c$. An expansion over a small number of vibrational states enables the determination of the amplitudes ratio for the open channels i.e. of the ratio between the AD cross sections ; the absolute values of the cross sections can be obtained through a normalization of the wavefunction using a large expansion. However, previous calculations[3] have shown that the associative detachment probability P_d is equal to unity, as soon as the system is able to enter into the small R region and that the size of the AD cross section is determined by the large R behaviour of the HCl^- potential energy curve. In the present work, we set the detachment probability to 1 inside a radius such that the detachment rate is equal to half the Langevin limit[5].

The boundary condition at $r = r_c$ is expanded to first order around the HCl equilibrium distance ; the corresponding two parameters were adjusted to reproduce the experimental results of Allan and Wong[1] on the temperature dependence of the DA cross sections. However the full optionization of the parameters has still to be performed, and we will only present preliminary results.

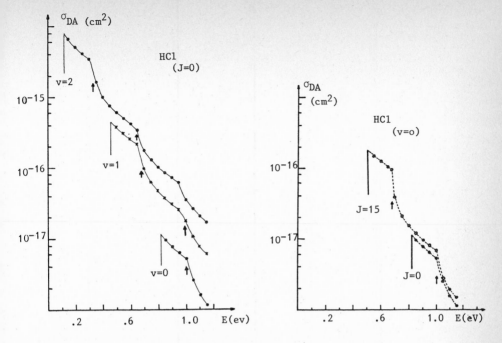

Fig. 1 : Dissociative attachment cross sections for e⁻-HCl collisions, for various initial vibrational states (v,J=0). The arrows indicate the location of the opening/closing of the v' vibrational states.

Fig. 2 : Dissociative attachment cross sections for e⁻-HCl collisions, for various initial rotational states (v=0, J=0 and 15)

The cross sections for electron attachment on various v and J levels are presented on figures 1 and 2. All of them display the same shape : a vertical onset at threshold followed by a rapid fall off. It is noteworthy, that the various (J) cross sections almost scale on a universal curve. Each curve is structured by steps at each opening/closing of a vibrational channel. These steps are easily understood if one looks at the associative detachment[3,6] process : detachment hugely favors the highest accessible vibrational level ; when the energy goes through a vibrational threshold, the predominantly populated v level changes, causing drastic changes (steps) in all channels[3]. The steps appear roughly for the same incoming electron energy, the small differences coming from the anharmonicity of the HCl molecule. These structures in the individual cross sections are reflected in the cross section for attachment on hot molecules : its structure comes from the product of the step structured individual cross sections by the exponentially decreasing thermal distribution. Figures 3,4 and 5 present a comparison between these cross sections with the experimental results of Allan and Wong[1] : As the temperature is increased, the cross section for attachment below the (v=0, J=0) threshold increases, with peaks showing up just below the opening of the v' vibrational channels.

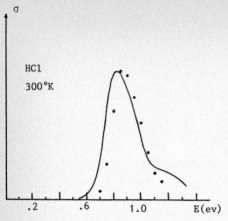 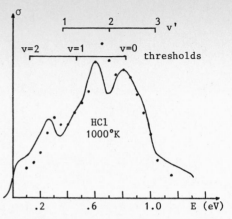

Figures 3 and 4 : Dissociative attachment cross sections for e⁻ thermal HCl collisions, at T=300° K and 1000°K. —— : experimental results of Allan and Wong[1] ; ● :
present results. The DA thresholds are indicated for the various v levels, as well
as the thresholds of the competitive vibrational channels (for v=0).

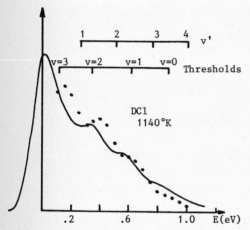

Fig. 5 : Same as figures 3 and 4,
for DCl targets at 1140°K

Although a complete optimization of the parameters has not been performed, the
general shape and temperature dependence of the experimental results are accounted
for by the present theory. This confirms the ability of the ERT formalism to treat
the interaction between electronic and vibrational motions with all dynamical
ouplings taken into account.

+(laboratoire associé au CNRS n° 281)

References

1. Allan M. and Wong S.F. 1981, J.Chem.Phys. 74 1687
2. Demkov Y.N. 1964 Sov.Phys. JETP 1964 19, 762 ; Dubé and Herzenberg 1977 Phys.Rev.
   Let. 38, 820 ; Fabrikant 1978, J.Phys.B 11, 3821 et 1983 16, 1253.
3. Gauyacq J.P. 1982 J.Phys.B 15, 2721
4. Gauyacq J.P. 1983 J.Phys.B in press
5. Howard C.J., Fehsenfeld F.C. and McFarland M. 1974, J.Chem.Phys. 68, 271-9
6. Gauyacq J.P. lecture notes in Chemistry, same volume
7. Herzenberg A. 1982, Notas de Fisica, 5, 225

RESONANCE-RESONANCE COUPLING IN DISSOCIATIVE ELECTRON ATTACHMENT

A. U. Hazi

Atomic and Molecular Theory Group

Lawrence Livermore National Laboratory

University of California, Livermore, California USA

Introduction

The effects of coupling between two adjacent electronic resonance states have been observed in several experiments involving dissociative attachment (DA) of electrons to diatomic molecules. When the potential energy curve of one resonance is attractive (possessing vibrational levels) and that of the other is repulsive, the coupling between the resonances may lead to the appearance of "vibrational" structure superimposed on a broad peak in the DA cross section. Such structure has been seen experimentally in H_2[1,2], HCl[3,4], HBr[3,5], HF[6], CO[7], NO[8] and some other molecules. When the potential energy curves of both resonances are repulsive, the effects of the coupling are less obvious. Nevertheless, such effects have been observed recently in experiments involving the formation of H^- by electron attachment to HBr[5].

The purpose of this paper is to discuss the effects of resonance-resonance coupling on dissociative electron attachment. I will briefly review some typical experimental results, and the two alternative theoretical interpretations[9-11] which have been proposed. I will present the main features of a theory[11] which involves indirect electronic coupling between the resonances and show how it can be applied to the formation of H^- by electron attachment to HBr. It appears that, in the case of hydrogen halides (HX), DA provides information about the __wavefunctions__ of the two HX^- resonances and the extent of spin-orbit mixing between them.

Experimental Results and Interpretations

Resonance-resonance coupling in DA was first observed by Dowell and Sharp[1] in experiments involving H^- formation from H_2 in the 10-13 eV region. They found a series of small peaks superimposed on a broad peak in the DA cross section. A subsequent, more detailed, experimental study[2] confirmed the earlier findings and showed that the energies of the "vibrational" peaks correspond to those of resonance series "a" which has been observed by several groups in both vibrational[12] and electronic excitation[13]. Figure 1 illustrates the experimental situation using the data of Tronc et al.[2]

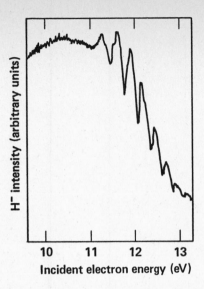

Fig. 1 Cross section for dissociative attachment of electrons to H_2 (from Ref. 2).

Taylor, Nazaroff and Golebiewski[9] were the first to suggest that these structures in the DA cross section of H_2 were due to the predissociation of the vibrational levels associated with an attractive resonance state of H_2^-, which was later identified to be the $(1\sigma_g 1\pi_u^2)$ $^2\Sigma_g^+$ state. The predissociation is caused by the repulsive $(1\sigma_g 1\sigma_u^2)$ $^2\Sigma_g^+$ state correlating with the $H+H^-$ dissociation limit asymptotically. This mechanism, which involves non-adiabatic coupling between the two $^2\Sigma_g^+$ resonances, is shown in Figure 2a. Also in 1966, O'Malley[10] proposed an alternative interpretation of the H_2 experiments[1] in terms of indirect electronic interaction between the resonances. This mechanism involves coupling via the electron scattering continuum associated with the vibrational levels of the ground state of H_2, as shown by the vertical broken arrows in Figure 2b. Since in this case the coupling involves the electronic Hamiltonian, the two resonances must have the same symmetry. This requirement is satisfied for H_2.

The case of two interacting resonances, each with a repulsive potential energy curve, is illustrated by H^- formation via DA to the hydrogen halide molecules. The dissociation limit $H^-(^1S) + X(^2P)$ gives rise to two resonances with $^2\Sigma^+$ and $^2\Pi$ symmetries in LS-coupling. However, spin-orbit coupling gives rise to actually three resonances: $^2\Sigma_{1/2}^+$, $^2\Pi_{3/2}$ and $^2\Pi_{1/2}$. In all cases, except HF^-, the $^2\Sigma^+$ resonance is lower in energy, so that the $^2\Sigma_{1/2}^+$ and $^2\Pi_{3/2}$ states dissociate adiabatically to the limit $H^- + X(^2P_{3/2})$, whereas the $^2\Pi_{1/2}$ state dissociates to $H^- + X(^2P_{1/2})$. The potential energy curves[5] appropriate for the case of HBr^- are shown in Figure 3.

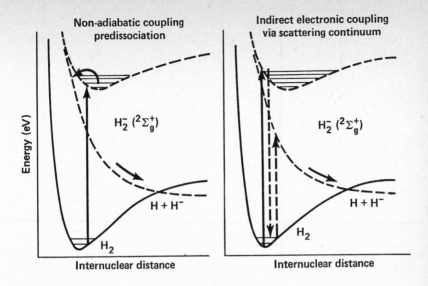

Fig. 2 Alternative theoretical interpretations of resonance-resonance coupling in H_2^-.

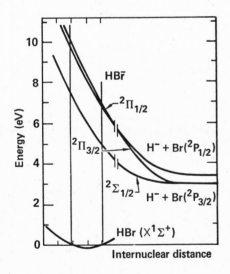

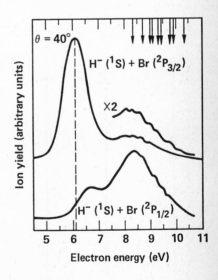

Fig. 3 Potential energy curves of HBr⁻ resonances (from Ref. 5). The low-energy $^2\Sigma^+$ resonance, disso-ciating to H+Br⁻, is not shown.

Fig. 4 Ion yield spectra for H⁻ formation by electron attachment to HBr (from Ref. 5).

In the heavier halogen atoms, the spin-orbit splitting of the $^2P_{1/2}$ and $^2P_{3/2}$ fine-structure levels is sufficiently large and it is possible to distinguish experimentally between the processes leading to the formation of $X(^2P_{3/2})$ and $X(^2P_{1/2})$. Based on the potential energy curves shown in Figure 3, one would expect two peaks, at ~6.1 eV and ~8.3 eV, in the ion yield curve for H^- produced from HBr^-. Furthermore, both the 6.1 eV and 8.3 eV peaks should appear in the H^- + Br $(^2P_{3/2})$ channel, while only the 8.3 eV peak is expected in the H^- + Br $(^2P_{1/2})$ channel. Figure 4 displays the actual experimental results of Le Coat, Azria and Tronc[5]. The most striking feature of Figure 4 is the appearance of an unexpected peak in the 5.5 – 7.5 eV region in the ion yield curve corresponding to the H^- + Br$(^2P_{1/2})$ channel. Le Coat et al.[5] attributed this observation to non-adiabatic coupling between the $^2\Sigma^+_{1/2}$ and $^2\Pi_{1/2}$ electronic states of HBr^-. However, a calculation using d/dR coupling gave too small an effect and did not account for the experimentally determined branching ratios[5]. Nevertheless, measurements of the angular distribution of H^- verified[5] that the 6.1 eV feature is indeed due to a $^2\Sigma^+$ resonance with a dominant $d\sigma$-wave character. Finally, the structure on the high-energy side of the 8.3 eV peak was attributed[5] to interaction between the repulsive $^2\Pi$ resonance and attractive Feshbach resonances observed previously in transmission experiments[14].

Theory of Electronic Coupling

The formal theory of electronic coupling between resonances has been presented elsewhere.[11] Here, I summarize only the most salient features of the theory. Qualitatively, this coupling can occur because states which can autodetach are not stationary states of the full electronic Hamiltonian, $H_{e\ell}$. In molecules, they are usually chosen to be adiabatic eigenstates of $QH_{e\ell}Q$, where Q is a suitable projection operator[10]. Consequently, the resonance states can interact with each other in second order, via the electron scattering continuum (described by $PH_{e\ell}P$) into which the resonances can decay. The formal theory[11] shows that the resonances are coupled by the same complex energy dependent optical potential which is responsible for the finite lifetimes of isolated resonances.

In the adiabatic nuclei approximation the total resonant wavefunction $Q\bar{\Phi}^+_E$ satisfies the equation

$$(E-K_R-QH_{e\ell}Q-QH_{e\ell}P\ G^+_{PP}(E)\ PH_{e\ell}Q)\ Q\bar{\Phi}^+_E = QH_{e\ell}P\phi^+_E \tag{1}$$

where K_R is the nuclear kinetic energy operator and $P = 1-Q$. The non-resonant scattering is described by the equation

$$(E-K_R-PH_{e\ell}P)P\phi^+_E = 0 \tag{2}$$

with the corresponding operator $G^+_{PP}(E)$ defined as

$$G^+_{PP}(E) = (E + i\epsilon - K_R - PH_{e\ell}P)^{-1} \tag{3}$$

For two interacting resonances, with electronic wave functions ϕ_i (i = 1,2), $Q\bar{\Phi}^+_E$ has the form

$$Q\Phi_E^+ = \phi_1\xi_1(R) + \phi_2\xi_2(R) \tag{4}$$

The nuclear wavefunctions $\xi_i(R)$ describe the motion of the nuclei in the compound states, and their asymptotic forms at large R give the DA amplitudes directly. They satisfy the coupled equations

$$(E-K_R-\varepsilon_i(R) - W_{ii})\xi_i - \sum_{j=i} W_{ij}\xi_j = U_i(R) \quad i,j = 1,2 \tag{5}$$

where $\varepsilon_i(R)$ denote the potential energy curves of the resonances and $U_i(R)$ are the entry amplitudes, i.e.,

$$\varepsilon_i(R) = \langle\phi_i H_{e\ell}\phi_i\rangle \qquad i = 1,2 \tag{6}$$

and

$$U_i(R) = \langle\phi_i H_{e\ell}P\phi_E^+\rangle \qquad i = 1,2 \tag{7}$$

The energy dependent matrix elements W_{ij} have the form

$$W_{ij}(R) = \langle\phi_i H_{e\ell}PG_{PP}^+(E) P H_{e\ell}\phi_j\rangle \tag{8}$$

It is clear from Eqns. (5) and (8) that the nuclear motions in the two electronic resonance states are coupled by W_{12}. This coupling results from the same energy dependent, complex potential which gives rise to the width Γ and the shift Δ of the resonances[10] (e.g., $W_{11} = \Delta_1 - i\Gamma_1/2$, etc.).

The coupled equations in (5) can be solved explicitly to yield the amplitude for DA in channel 2:

$$T_2^{DA} = \langle y_2^- U_2\rangle + \langle y_2^- W_{21} \frac{1}{E-K_R-\varepsilon_1-W_{11}} U_1\rangle \tag{9}$$

Here the nuclear wavefunction $y_2^-(R)$ describes dissociation along the complex potential $(\varepsilon_2 + W_{22})$, i.e., it satisfies

$$(E-K_R-\varepsilon_2-W_{22})y_2^- = 0 \tag{10}$$

with incoming boundary condition. The first term on the right hand side of Eq. (9) represents the customary, direct dissociation via resonance 2. The second term represents the indirect mechanism in which the electron is captured into resonance 1, with the nuclei moving in the complex potential (ε_1+W_{11}), and then the electronic coupling W_{21} induces a transition to resonance 2 which dissociates. Since the DA cross section is proportional to $|T_2^{DA}|^2$, the direct and indirect mechanisms interfere with each other. Furthermore, it is clear from Eq. (9) that the cross section can exhibit structure if resonance 1 is attractive because the potential $(\varepsilon_1 + W_{11})$ has bound vibrational states which, in turn, give rise to complex poles in the indirect amplitude in Eq. (9). This is the situation illustrated by Figure 1 and Figure 2b for H_2.

Application to HX⁻

In order to apply the formalism of the previous section to the process of H^- formation by electron attachment to hydrogen halides, it is necessary to consider the effects of the spin-orbit interaction that mixes the ${}^2\Sigma_{1/2}^+$ and

$^2\Pi_{1/2}$ electronic resonances. In this case, the resonance wavefunctions can be written as

$$\phi_1 = (1-c^2)^{1/2} \, \phi_\Sigma + c \, \phi_\Pi \tag{11a}$$

$$\phi_2 = -c\phi_\Sigma + (1-c^2)^{1/2}\phi_\Pi \tag{11b}$$

where c is the spin-orbit mixing coefficients. Assuming that the so-called "local approximation" is valid[15] and that the shifts are negligible, one obtains, to first order in c:

$$W_{11} = -i/2 \, \Gamma_\Sigma, \qquad W_{22} = -i/2\Gamma_\Pi \tag{12a}$$

and

$$W_{12} = W_{21} = -i/2c(\Gamma_\Pi - \Gamma_\Sigma) \tag{12b}$$

Thus, in this approximation, the interaction between the two resonances is proportional to the spin-orbit mixing coefficient and to the difference between the widths of the two resonances. Of course, all these quantities are functions of the internuclear distance. This result suggests that, in the hydrogen halides, the DA cross section provides some information about not only the usual resonance parameters (ϵ and Γ) but also the electronic wavefunctions of the compound states involved.

Future Work

Quantitative verification of the indirect electronic coupling mechanism must await detailed calculations of the resonance parameters of the $^2\Sigma^+$ and $^2\Pi$ resonances of HX$^-$ and of the spin-orbit mixing between the $\Omega = 1/2$ components. Electronic coupling between resonances will also have an effect on the angular distribution of the fragments. A theoretical study of the differential cross section for H$^-$ formation from HX molecules is in progress.[16] The expected interference between the direct and the indirect mechanisms of DA (see Eq. (9)) is another intriguing effect. It is not unreasonable to suggest that the observed[5] shift of the $^2\Sigma^+$ peak from 6.1 to 6.6 eV in the ion yield spectrum of the H$^-$+Br ($^2P_{1/2}$) channel (see Figure 4) is due to this interference. In addition, the detailed shapes of the "vibrational" structures seen in DA cross sections (see Figures 1 and 4) are expected to be quite sensitive to the relative phases of the direct and indirect DA amplitudes. This feature of the theory also needs further investigation.

Acknowledgement

Work performed under the auspices of the U. S. Department of Energy by the Lawrence Livermore National Laboratory under Contract No. W-7405-ENG-48.

References

1. J. T. Dowell and T. E. Sharp, Phys. Rev. <u>167</u>. 124 (1968).
2. M. Tronc, F. Fiquet-Fayard, C. Schermann, and R. Hall, J. Phys. B <u>10</u>, 305 (1977).
3. J. P. Ziesel, I. Nenner and G. J. Schulz, J. Chem. Phys. <u>63</u>, 1943 (1975).
4. R. Azria, Y. LeCoat, D. Simon and M. Tronc, J. Phys. B <u>13</u>, 1909 (1980).
5. Y. Le Coat, R. Azria, and M. Tronc, J. Phys. B <u>15</u>, 1569 (1982).
6. R. Abouaf and D. Teillet-Billy, J. Phys. B <u>13</u>, L275 (1980).
7. A. Stamatovic and G. J. Schulz, Rev. Sci. Instrum. <u>41</u>, 423 (1970).
8. D. Rapp and D. D. Briglia, J. Chem. Phys. <u>43</u>, 1480 (1965); P. J. Chantry, Phys. Rev. <u>172</u>, 125 (1968), C. Paquet, P. Marchand and P. Marmet, Can. J. Phys. <u>49</u>, 2013 (1971).
9. H. S. Taylor, G. V. Nazaroff and A. Golebiewski, J. Chem. Phys. <u>45</u>, 2872 (1966), H. S. Taylor, Adv. Chem. Phys. <u>18</u>, 91 (1970).
10. T. F. O'Malley, Phys. Rev. <u>150</u>, 14 (1966).
11. A. U. Hazi, J. Phys. B <u>16</u> L29 (1983).
12. J. Comer and F. H. Read, J. Phys. B <u>4</u>, 368 (1971).
13. A. Weingartshofer, H. Ehrhardt, V. Hermann and F. Linder, Phys. Rev. A <u>2</u> 294 (1970); S. B. Elston, S. A. Lawton and M. Pichanik, Phys. Rev. <u>A 10</u>, 225 (1974).
14. D. Spence and T. Noguchi, J. Chem. Phys. <u>63</u>, 505 (1975).
15. A. U. Hazi, T. N. Rescigno and M. Kurilla, Phys. Rev. A <u>23</u>, 1089 (1981).
16. A. Hazi, unpublished.

CALCULATION OF CROSS SECTIONS FOR DISSOCIATIVE ATTACHMENT AND VIBRATIONAL EXCITATION IN HCl BEYOND THE LOCAL COMPLEX POTENTIAL APPROXIMATION

W. Domcke and C. Mündel

Theoretische Chemie, Physikalisch-Chemisches Institut,
Universität Heidelberg, D-69 Heidelberg, West Germany

The general theory of dissociative attachment, vibrational excitation by electron impact, and related reaction processes has been developed by Chen [1], O'Malley [2], Bardsley [3] and others within the framework of Feshbach's projection operator formalism. In this formulation, the nuclear dynamics in the negative ion is governed by an "optical" potential V_{opt} which may be written as

$$V_{opt} = V_1(R) + \Delta(R,E-H_o) - \frac{1}{2}i\Gamma(R,E-H_o) \tag{1}$$

where

$$H_o = T_N + V_o(R) \tag{2}$$

$$\Gamma(R,E) = 2\pi|v_E(R)|^2 \tag{3}$$

$$\Delta(R,E) = \frac{P}{2\pi} \int dE'\,\Gamma(R,E')/(E-E') \tag{4}$$

Here E is the total energy, R the internuclear distance, T_N the nuclear kinetic energy operator, $V_o(R)$ the potential energy curve of the target molecule, and $V_1(R)$ the potential energy curve of a suitable diabatic discrete electronic state of the negative ion [1-3]. The discrete state is coupled via $v_E(R)$ to the electronic continuum of the target molecule, which results in a decay width Γ and a level-shift Δ as a given in eqs. (1-4).

The effective potential (1) is energy-dependent, complex and non-local. The non-locality results from the explicit dependence of the width Γ and the level-shift Δ on the nuclear kinetic energy operator T_N and represents thus a non-Born-Oppenheimer effect. To simplify the treatment of the nuclear dynamics, V_{opt} is usually approximated by a local and energy-independent complex potential [1-4]. While the local approximation may be an excellent approximation for resonances sufficiently far from threshold [5], it is expected to be poor for the treatment of dissociative attachment and associative detachment, where the resonance crosses the threshold as a function of the internuclear distance [3,6].

The cross section for dissociative attachment as a function of the energy E_i of the incident electron may be written as

$$\sigma_{diss}(E_i) = \frac{\pi}{2E_i}\frac{\mu}{K}|<0|v_{E_i}|K^{(+)}>|^2 \tag{5}$$

where $|0>$ is the vibrational ground state of the target molecule and $|K^{(+)}>$ is the ion-atom scattering state in the effective potential V_{opt}. μ is the reduced mass of the fragments in the dissociation channel and K the corresponding asymptotic momentum. The angle-integrated cross section for $0 \rightarrow v$ vibrational excitation reads

$$\sigma_{vo}(E_i) = \frac{\pi}{2E_i}|<v|v_{E_f} G v_{E_i}^*|0>|^2 \tag{6}$$

where E_f is the final kinetic energy of the inelastically scattered electron and $G = (E-T_N-V_{opt})^{-1}$.

Our method to compute $|K^{(+)}>$ and G beyond the local approximation is based on the parametrization of $V_o(R)$ and $V_1(R)$ by Morse potentials. The rotational motion is neglected, which is a good approximation for electron scattering and dissociative attachment in low-temperature gases. The bound states, scattering states and the Green's function of the non-rotating Morse oscillator are known analytically. This allows us to treat the strongly repulsive local part $V_1(R)$ of the effective potential V_{opt} exactly. The non-local part $F = \Delta - 1/2\ i\Gamma$ is approximated by the separable expansion

$$F^{(s)} = \sum_{\nu,\mu=1}^{N} F|\chi_\nu>(f^{-1})_{\nu\mu}<\chi_\mu|F, \qquad (7)$$

$$f_{\nu\mu} = <\chi_\nu|F|\chi_\mu>$$

where $<R|\chi_\nu>$ are square-integrable basis functions. This expansion reduces the Lippmann-Schwinger equation for $|K^{(+)}>$ to a finite set of linear algebraic equations /7/. A suitable complete orthonormal set of L^2 basis functions is provided by the Lanczos basis /8/ of the Morse Hamiltonian $H_1 = T_N + V_1(R)$. The completeness of the Lanczos basis guarantees the convergence of the expansion (7). For more details, the reader is referred to ref. 9.

Based on this formalism, we have performed model calculations for vibrational excitation and dissociative attachment in HCl. The model is specified by the functions $V_o(R)$, $V_1(R)$ and $\Gamma(R,E)$. The Morse parametrization of the $^1\Sigma^+$ ground state potential energy curve of HCl is well known from the analysis of spectroscopic data. The width function has been parametrized as

$$\Gamma(R,E) = f(R)\ E^\alpha\ e^{-\beta E} \qquad (8)$$

where $f(R)$ is a Gaussian and $\alpha = 0.36$ is the threshold exponent /10/ corresponding to the dipole moment of HCl at the equilibrium geometry. $\Delta(R,E)$ is obtained as the Hilbert transform of $\Gamma(R,E)$ according to eq. (4). The parameters characterizing the (repulsive) Morse potential $V_1(R)$ and $\Gamma(R,E)$ have been determined as follows. First, the parameters were varied until the fixed-nuclei eigenphase sum $\delta(R,E)$ of the model was in qualitative agreement with the ab initio calculations of Padial, Norcross and Collins /11/ for the $^2\Sigma$ eigenphase sum for e-HCl scattering. A final slight adjustment of the parameters was then made to improve the agreement of the calculated cross sections with the experimental data.

Fig. 1 shows the potential energy curves $V_o(R)$ (long dashes), $V_1(R)$ (short dashes) and the width and level-shift functions $\Gamma(R,E)$, $\Delta(R,E)$ of the present model. The adiabatic potential energy curve $V_{res}(R)$ of the resonance and bound state of HCl⁻ is also shown in fig. 1a (full line). The singular behaviour of $\Delta(E)$ at $E = 0$, which is caused by the long-range dipole potential, results in an rather unusual shape of the adiabatic potential curve in the crossing region which has been discussed in detail previously /10/. The multi-valuedness of $V_{res}(R)$ clearly excludes the treatment of the nuclear dynamics in the local approximation.

The calculated exact (within the model considered) integral cross sections for $v = 0 \to 1$ and $v = 0 \to 2$ vibrational excitation are shown in fig. 2. The cross sections exhibit a broad resonance peak near 3.5 eV and a strong and narrow peak at threshold. Both the shape of the cross sections as well as their absolute magnitude is in semiquantitative agreement with the measurements of Rohr and Linder /12/. The broad resonance peak corresponds to a shape resonance in the $^2\Sigma$ symmetry /11/, while the narrow peaks represent a dipole-induced threshold effect.

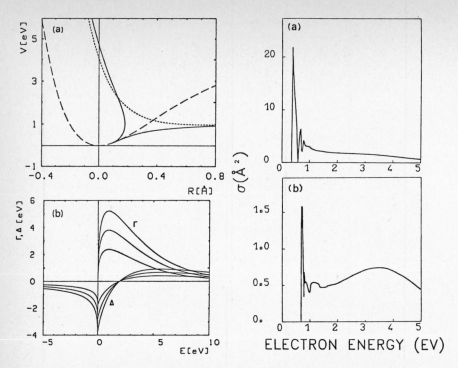

Fig. 1: Potential energy curves (a) and width and level-shift functions (b) of the present e-HCl scattering model.

Fig. 2: Calculated integral cross sections for 0→1 (a) and 0→2 (b) vibrational excitation in HCl.

Fig. 3 shows the calculated cross section for dissociative attachment in comparison with recent absolute measurements /13/. There is good overall agreement, both in shape as well as in absolute magnitude. In particular, the characteristic step structure /14/ in the dissociative attachment cross section at the threshold energies of the vibrational excitation channels of HCl is reproduced by the calculation.

The present calculations represent the first fully non-local treatment of the dissociative attachment process in a polar molecule. In contrast to previous attempts /15/ we have taken account of the non-local level-shift operator $\Delta(R, E-H_o)$, which is essential to reproduce the threshold peaks in vibrational excitation of HCl.

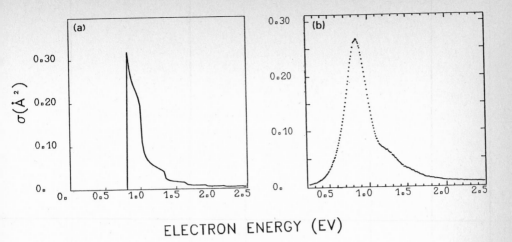

ELECTRON ENERGY (EV)

Fig. 3: Calculated (a) and experimental /13/ (b) cross section for
dissociative attachment in HCl.

References

1. J.C.Y. Chen, Phys. Rev. 148, 66 (1966)
2. T.F. O'Malley, Phys. Rev. 150, 14 (1966)
3. J.N. Bardsley, J. Phys. B 1, 349, 365 (1968)
4. A. Herzenberg, J. Phys. B 1, 548 (1968)
5. M. Berman, H. Estrada, L.S. Cederbaum and W. Domcke, Phys. Rev. A 28, 1363 (1983)
6. R.J. Bieniek, J. Phys. B 13, 4405 (1980); A.U. Hazi, A.E. Orel and T.N. Rescigno,
   Phys. Rev. Lett. 46, 918 (1981)
7. S.K. Adhikari and I.H. Sloan, Phys. Rev. C 11, 1133 (1975)
8. C. Lanczos, J. Res. Natl. Bur. Stand. 45, 367 (1950)
9. C. Mündel and W. Domcke, to be published
10. W. Domcke, J. Phys. B 14, 4889 (1981); W. Domcke and L.S. Cederbaum, J. Phys. B
    14, 149 (1981)
11. D.W. Norcross, private communication
12. K. Rohr and F. Linder, J. Phys. B 9, 2521 (1976)
13. O.J. Orient and S.K. Srivastava, to be published
14. R. Abouaf and D. Teillet-Billy, J. Phys. B 10, 2261 (1977)
15. F. Fiquet-Fayard, J. Phys. B 7, 810 (1974); J.N. Bardley and J.M. Wadehra,
    J. Chem. Phys. 78, 7227 (1983)

PART IV

SCATTERING OF HIGH-ENERGY

ELECTRONS BY ATOMS AND

MOLECULES: THE EXPERIMENTAL STUDIES

DIPOLAR ELECTRON MOLECULE EXPERIMENTS

C.E. Brion
Department of Chemistry
University of British Columbia
Vancouver, B.C. V6T 1Y6
Canada

ABSTRACT

The current status of fast electron experiments simulating photoabsorption and photoionization techniques with tuneable energy transfer is reviewed. Recent results for "photoabsorption" (e,e), "photoelectron spectroscopy" (e,2e) and "photoionization mass spectrometry" (e,e + ion) are presented. Suggestions are made for new highly sensitive spectrometers for the study of electron-molecule dipolar collisions. A new oscillator strength compilation is discussed.

INTRODUCTION

Although tuneable energy synchrotron radiation has become increasingly available in recent years for direct photoabsorption and photoionization experiments, gas phase work has been quite restricted in scope and many such studies have been limited by considerations of intensity, resolution and spectral range mainly as a result of the properties of the dispersing monochromators. In particular absolute partial oscillator strengths (cross-sections) for photoionization are difficult to measure. The demand for this type of data has greatly increased also with a view to understanding the dipole breakdown of molecules and for the evaluation of new types of quantum calculation methods and continuum wave functions [44].

With the foregoing considerations in mind we started in the early 1970's, at the University of British Columbia, to develop alternative new methods of measuring abso-lute oscillator strengths for photoabsorption and photoionization in both total and partial channels. All these new methods exploit the virtual photon field of a fast electron which induces a dipole field in the target species. The principles and underlying scattering theory have been reviewed in earlier publications [1,2,3] and only brief mention of these will be made here. Some of the early work [1] was done in collaboration with M.J. Van der Wiel and co-workers at the FOM Institute in Amsterdam.

In the limit of vanishingly small momentum transfer (achieved by using electrons of several kiloelectron volts energy and zero degree scattering angle) the Bethe-Born theory, as discussed by Lassettre [4] and Inokuti [5] relates the dipole (or optical) oscillator strength $\left(\frac{dfo}{dE}\right)$ to the differential electron scattering cross-section $\left(\frac{d\sigma}{dE}\right)$ according to the equation

$$\frac{dfo}{dE} = \frac{E}{2} \frac{ko}{kn} K^2 \frac{d\sigma}{dE} \qquad (1)$$

where E is the electron energy loss (analogous to the photon energy) and ko, kn and K are the incident, scattered and transfer momenta respectively. Thus dipole oscillator strengths may be obtained directly by simple kinematic conversion of the electron scattering intensities. In practice, of course, the scattering intensities are only determinable on a relative basis. However unlike most photon continua the virtual photon field has equal intensity at all energy transfers (E) used [1,2] and thus the correct relative shapes of dipole oscillator strength spectra are obtained. It

has now been repeatedly and convincingly shown that the wide range, relative photoabsorption spectrum thus obtained can readily be put on an absolute scale by sum rule
normalization [1,2]. Specifically the TRK sum rule equates the total dipole oscillator strength to the number of electrons in the target species. In practice an effective shell separation exists for many molecules and the valence shell spectrum can be
normalized to the number of valence electrons with a small correction for Pauli
excluded transitions [6]. In this way absolute oscillator strengths with an accuracy
better than 5% are obtained directly without making any absolute measurements. This
is a very attractive feature of the electron impact simulation methods.

The application of the above ideas in conjunction with the use of electron
energy loss spectroscopy (in some cases also with coincidence counting) has resulted
in the measurement of a large number of dipole oscillator strengths for (1) Photoabsorption, by dipole (e,e) spectroscopy, (2) Partial photoionization to electronic
states, by dipole (e,2e) spectroscopy (simulating tuneable energy photoelectron
spectroscopy) and (3) Molecular and dissociative photoionization by dipole (e,e +
ion) spectroscopy (simulating tuneable energy photoionization mass spectrometry).
These methods have been fully discussed in earlier articles [1,3, 7-10]. In addition the general basic theory of the dipole (e,2e) method has been investigated [11]
while recently a more detailed and elegant theoretical treatment has been given by
White [12].

The effectiveness and validity of these photon simulation experiments is
clearly demonstrated by comparing with directly obtained photoabsorption and photoionization data in those cases where the direct optical studies have been made. Some
examples are for photoabsorption of N_2O [13,14] and COS [15,16] as well as partial
photoionization of H_2O [17,18] and $CO\overset{+}{S}$ [15,16] and the dissociative photoionization
of O_2 [19,20].

Using the dipole electron molecule technique optical oscillator strengths have
been measured for a wide variety of photon induced processes for a large range of
target molecules. These published results have now been incorporated into a generalized data compilation [21] to be discussed in a later part of the present article.

In the following sections recent advances in and the current status of the
instrumentation are discussed together with some new possibilities promising increased resolution and sensitivity for dipole electron molecule experiments. Finally
the data compilation is reviewed.

INSTRUMENTATION

(a) Dipole (e,e) Spectrometers. Our existing high resolution spectrometer [22] for
valence and inner shell electronic excitation spectra is continuing to produce good
results. However it suffers from several disadvantages. Firstly since it is contained entirely within a single vacuum chamber it is rapidly contaminated by reactive
gases, particularly by those that decompose on the hot tungsten cathode. This
results in rapid deterioration of both sensitivity and resolution and necessitates
frequent tedious dismantling and cleaning procedures. In the case of inner shell
spectra [22] small angle scattering must be employed to reduce the large non-spectral
backgrounds that occur when the primary electron beam is allowed to enter the
analyser. In order to overcome these operational difficulties and provide improved
overall performance a new high resolution (e,e) spectrometer (SUPERSPEC) has been
designed, built and tested [3,23]. This new instrument, which also affords large
increases in sensitivity, is shown in figure 1. The design features and operation
have been fully discussed in a recent publication [23]. The large hemispherical
electron analysers (mean diameter 16 inches) and sophisticated electron optics ensure
high sensitivity and high resolution at high impact energy while maintaining high
analyser pass energies and reasonable lens ratios. The instrument functions well
[3,23] at zero degrees scattering angle with negligible non-spectral background for
both valence and inner shell spectra. This much improved performance can be seen

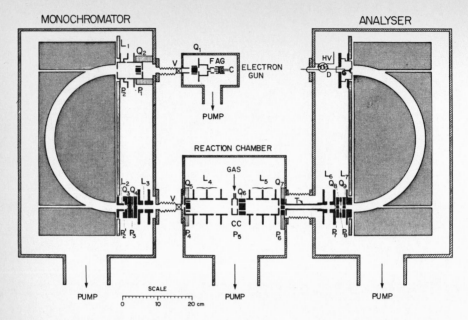

MONOCHROMATOR

ANALYSER

ELECTRON GUN

PUMP

REACTION CHAMBER

GAS

CC

PUMP

PUMP

PUMP

SCALE

0   10   20 cm

Figure 1

from the spectra in figures 2 and 3. Most notable is the vibrationally resolved high resolution K-shell spectrum of the ls → π* transition of N_2 (figure 3) obtained at zero degrees in one hour collection time compared with the several days needed for earlier measurements [24,25] with similar statistics. Figure 4 shows a low resolution, wide range K-shell spectrum of N_2 obtained in 10 minutes. The spectrometer is presently being fitted with a microchannel plate position sensitive detector which should result in further large improvements in performance. This detector is similar

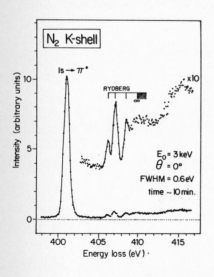

N_2 K-shell

ls → π*

RYDBERG

∞

x10

E_0 = 3 keV
θ = 0°
FWHM = 0.6 eV
time ~ 10 min.

Intensity (arbitrary units)

Energy loss (eV)

Figure 2

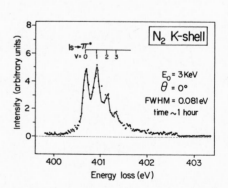

N_2 K-shell

ls → π*
v = 0   1   2   3

E_0 = 3 KeV
θ = 0°
FWHM = 0.081 eV
time ~ 1 hour

Intensity (arbitrary units)

Energy loss (eV)

Figure 3

to that described by Hicks et al [26] but uses fibre optics to couple directly the output of a phosphor screen to a photodiode array. The spectrometer is differentially pumped with separate sections for the gun, monochromator, collision chamber and analyser regions. This permits study of reactive gases and eliminates almost all need for cleaning. Furthermore, neither the spectrometer tuning nor the absolute energy loss scale are affected by gas introduction or change of gas samples. The zero degree scattering angle will permit future measurements of dipole oscillator strengths for both valence and inner shell processes. The increased sensitivity and differential pumping make the instrument ideal for the study of absolute oscillator strengths for atoms, radicals and ions, as well as transient and reactive species.

(b) Dipole (e,2e) Spectrometer. Full details of the design and operation of the existing magic angle coincidence spectrometer have been published earlier [7,15]. Future improvements are planned to increase both the sensitivity and resolution (at present 1.5 eV FWHM) of this instrument since both factors currently limit the range of possible studies. Sensitivity can be significantly improved by placing a micro-channel plate position sensitive detector in the focal plane of the ejected (54°) analyser. Coincidence detection with positional informa-tion can be obtained by taking the fast timing signal off the exit face of the channel plate. Recently Cook and Weigold [27] have convincingly demonstrated the effectiveness of a similar position sensitive coincidence system in a binary (e,2e) spectrometer. Following a recent suggestion by Cook [28] we are also exploring the possibility of obtaining high resolution (perhaps as good as 0.03ev) dipole (e,2e) spectra without any electron beam monochromation. This would be achieved by using the ideas put forward and experimentally demonstrated by Zscheile [29] who has shown how a thermal electron beam can be spatially dispersed into a parallel beam by an analyser without an exit slit. Electrons from this dispersed beam are then inelastically scattered off a target in the forward direction, analysed and all electrons with the same energy loss regardless of position are focussed at a single point in the exit plane of the forward analyser. Now if a microchannel plate is put in the exit plane of the ejected electron analyser of the dipole (e,2e) spectrometer it would be possible to collect a range of energy losses simultaneously. The successful combined use of channel plates in both the ejected and the forward analyser together with the idea of Zscheile [29] offers the possibility of very significant gains in instrumental sensitivity and resolution.

(c) Dipole (e,e + ion) Spectrometer. This instrument, originally built and operated by Van der Wiel et al [8,9] at the FOM Institute, Amsterdam, was moved in 1980 to the University of British Columbia. It has since undergone a number of minor improvements and modifications [10]. Resolution and sensitivity could both be improved using the ideas discussed in the preceding section on the dipole (e,2e) spectrometer. However on a relative basis the existing (e,e+ion) machine is already much more sensitive than the (e,2e) since the former collects all ions regardless of direction (and independent of the kinetic energy of fragmentation below 20eV) whereas the latter instrument is differential in that it only collects ejected electrons in a narrow cone about the magic angle of 54°. There is thus less need for improving sensitivity, particularly as present data rates are quite high in most cases.

RECENT RESULTS

(a) Dipole (e,2e) Spectroscopy. Valence and inner shell spectra for NF_3 [30], amines [31], PF_3 [32], $POCl_3$ [33] as well as allene and butadiene [34] and tetra-methyl silane [35] have been obtained recently using the original spectrometer [22]. A recent innovation has been the ability to obtain high resolution absolute oscillator strengths for valence shell photoabsorption in the discrete region. Results for HF [36] are shown in figure 4 and the oscillator strength scale was obtained by normalization on earlier low resolution (e,e) spectra [37] in the smooth continuum region. The spectrum has been assigned [31] on the basis of detailed quantum calculations of final states as well as computed oscillator strengths. A further recent development has been the accurate measurement of a series of inner shell transitions between 150 and 900ev for use as calibration standards in ISEELS and related studies

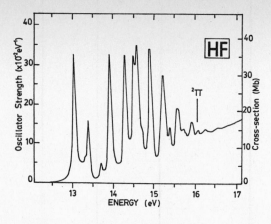

Figure 4

[38]. The new high performance (e,e) spectrometer [23] has been used for measuring high resolution core and valence spectra of HBr and HCl [39].

(b) <u>Dipole (e,2e) and (e,e+ion) Spectroscopies.</u> Results from both experiments will be discussed together due to their inter-relation. Following recent studies of the partial photoionization of HF to electronic states of the ion [40] as well as molecular and dissociative photoionization [10] we have now completed similar measurements for the related molecule HCl [41]. Figure 5 shows the absolute photoabsorption oscillator strength of HCl which has not been previously measured. The total oscillator strength is very small above 35eV and this results in poor statistics for the partial channel measurements at higher energies. Binding energy spectra shown in figure 6 are notable for the split ionic pole strength for 4σ ionization due to many-body effects. These latter effects for HCl have been studied in detail by binary (e,2e) spectroscopy [42] with all the structure above 20eV being assigned conclusively to the $(4\sigma)^{-1}$ process. This represents a breakdown of simple MO theory in the case of ionization of the inner valence orbital and is indicative of extensive electron correlation effects. Figure 7 shows the partial oscillator strengths for photoionization of the three valence orbitals up to 40eV while in figure 8 the results for the lowest two states of HCl^+ are compared with the recent calculations

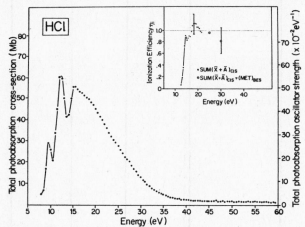

Figure 5

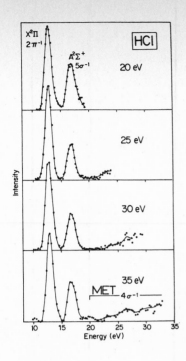

Figure 6

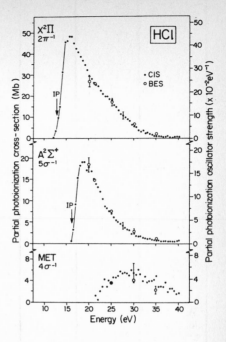

Figure 7

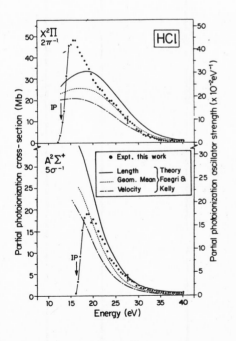

Figure 8

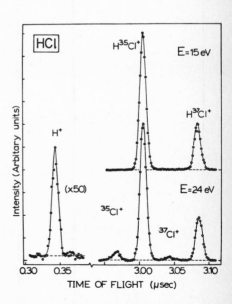

Figure 9

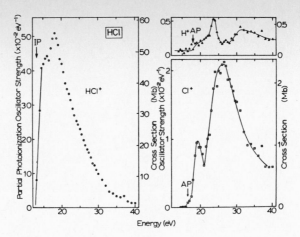

Figure 10

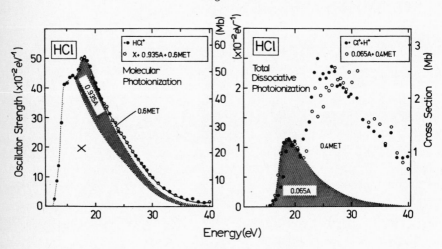

Figure 11

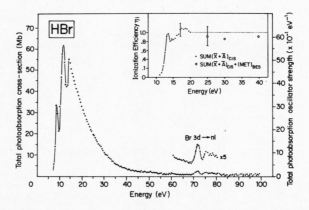

Figure 12

149

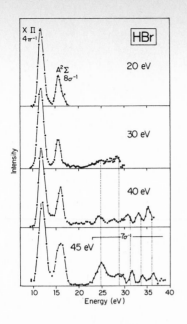

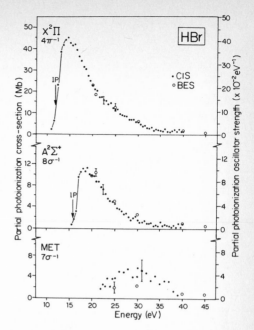

Figure 13   Figure 14

by Faegri and Kelly [43]. In figures 9 and 10 are shown typical time of flight mass
spectra and the partial oscillator strengths for molecular and dissociative photoionization of HCl. Combining the results of the dipole (e,2e) and (e,e+ion)
experiments a detailed dipole breakdown picture of HCl is obtained. Figure 11 shows
the respective electronic state contributions to the separate channels of molecular
and dissociative photoionization. Full details have been published elsewhere [41].
Figures 12, 13 and 14 show recent dipole (e,2e) results for HBr [44]. Dipole
(e,e+ion) results for NO have also been obtained [45].

OSCILLATOR STRENGTH COMPILATION

In the past decade a large number of valence shell optical oscillator strength
measurements have been made using the dipole electron molecule techniques discussed
above. Absolute oscillator strength measurements of photo-absorptioin, photoionization and ionic fragmentation for H_2, CO, N_2, O_2, NO, HF, HCl, HBr, H_2O, NH_3, CH_4,
N_2O, CO_2, COS, CS_2 and SF_6 over a wide energy range have now been collected together
and presented in standardized tables which are to be published [46]. A second
compilation [47] of molecular photoelectron branching ratios is also available.

ACKNOWLEDGEMENTS

This work received financial support from The Natural Sciences and Engineering
Research Council of Canada and also The Petroleum Research Fund administered by the
American Chemical Society. I am indebted to Dr. J.P.D. Cook for suggestions and
valuable discussions of ways to improve the performance of dipole (e,2e)
spectrometers. Finally I would like to thank my co-workers particularly S. Daviel,
Y. Iida, F. Carnovale, P. Thomson, R. Sodhi and A. Hitchcock who have been involved
with much of the new experimental work reported here.

REFERENCES
[1] C.E. Brion and A. Hamnett, in "The Excited State in Chemical Physics, Part 2",
Adv. Chem. Physics, volume 45, 1, (Ed. J.W. McGowan) John Wiley New York
(1981).
[2] C.E. Brion, in Physics of Electronic and Atomic Collisions. Ed. S. Datz (North-
Holland 1982) page 579.
[3] C.E. Brion, S. Daviel, R. Sodhi and A.P. Hitchcock, AIP Conference Proceedings
No. 94 - X-Ray and Atomic Inner-Shell Physics (Ed. B. Crasemann), American
Institute of Physics, (1982) page 429.
[4] E.N. Lassettre, Rad. Research (Supp) 1, 530 (1959).
[5] M. Inokuti, Rev. Mod. Phys. 43, 297 (1971).
[6] J.A. Wheeler and J.A. Bearden, Phys. Rev. 46, 755 (1934).
[7] C.E. Brion, A. Hamnett, G.R. Wight and M.J. Van der Wiel, J. Electron
Spectrosc., 12, 323 (1977).
[8] C. Backx and M.J. Van der Wiel, J. Phjys. B. 8, 3020 (1975).
[9] C. Backx, G.R. Wight, R.R. Tol and M.J. Van der Wiel, J. Phys. B. 8, 3007
(1975).
[10] F. Carnovale and C.E. Brion, Chem. Physics, 74, 253 (1983).
[11] A. Hamnett, W. Stoll, G. Branton, M.J. Van der Wiel and C.E. Brion, J. Phys.
B., 9, 945 (1976).
[12] M.G. White, Phys. Rev. A 26, 1907 (1982).
[13] C.E. Brion and K.H. Tan, Chem. Physics 34, 141 (1978).
[14] L.C. Lee, R.W. Carlson, D.L. Judge and M. Ogawa, J. Quant. Spectry. Rad.
Transfer 13, 1023 (1973).
[15] M.G. White, K.T. Leung and C.E. Brion, J. Electron Spectrosc., 23, 127 (1981).
[16] T.A. Carlson, M.O. Krause and F.A. Grimm, J. Chem. Phys. 77, 1701 (1982).
[17] K.H. Tan, C.E. Brion, Ph.E. Van er Leeuw and M.J. Van der Wiel, Chem. Physics,
29, 299 (1978).
[18] C.M. Truesdale, S. Southworth, P.H. Kobrin, D.W. Lindle, G. Thornton and D.A.
Shirley, J. Chem. Phys. 76, 860 (1982).
[19] C.E. Brion, K.H. Tan, M.J. Van der Wiel and Ph.E. Van der Leeuw, J. Electron
Spectrosc. 17, 101 (1979).
[20] J.A.R. Samson, G.H. Rayborn and P.N. Pareek, J. Chem. Phys. 76, 393 (1982).
[21] C.E. Brion and J.P. Thomson, to be published.
[22] A.P. Hitchcock and C.E. Brion, J. Electron Spectrosc., 13, 193 (1978).
[23] S. Daviel, C.E. Brion and A.P. Hitchcock, Rev. Sci. Instrum., submitted (1983).
[24] G.C. King, F.H. Read and M. Tronc, Chem. Phys. Letters 52; 50 (1977).
[25] A.P. Hitchcock and C.E. Brion, J. Electron Spectrosc. 18 (1980).
[26] P.J. Hicks, S. Daviel, B. Wallbank and J. Comer, J. Phys. E. Sci. Instrum. 13,
713 (1980).
[27] J.P.D. Cook and E. Weigold, to be published.
[28] J.P.D. Cook, private communication.
[29] H. Zscheile, J. Phys. E. Sci. Instrum. 15, 749 (1982).
[30] R.N. Sodhi, C.E. Brion and R.G. Cavell, to be published.
[31] R.N. Sodhi and C.E. Brion, to be published.
[32] R.N. Sodhi and C.E. Brion, to be published.
[33] C.E. Brion, R.N. Sodhi, S. Daviel and G.G.B. de Souza, to be published.
[34] A.P. Hitchcock, G.R.J. Williams, C.E. Brion and P.W. Langhoff, Chem. Physics,
submitted 1983.
[35] F.C. Carnovale, R. Tseng and C.E. Brion, J. Phys. B, 14, 4771 (1981).
[36] C.E. Brion and R.N. Sodhi, J. Electron Spectrosc., in press (1983).
[37] S. Daviel and C.E. Brion, to be published.
[38] S. Daviel, Y. Iida, F. Carnovale and C.E. Brion, Chem. Phys., in press (1983).
[39] C.E. Brion, S.T. Hood, I.H. Suzuki, E. Weigold and G.R.J. Williams, J. Electron
Spectrosc., 21, 71-91 (1980).
[40] K. Faegri Jr. and H.P. Kelly, Chem. Phys. Letters, 85, 472 (1982).
[41] C.E. Brion, Y. Iida and F. Carnovale, to be published.
[42] Y. Iida, S. Daviel, F. Carnovale and C.E. Brion, to be published.
[43] C.E. Brion and J.P. Thomson, to be published.
[44] G.H.F. Diercksen, W.P. Kraemer, T.N. Rescigno, C.F. Bender, B.V. McKay, S.R.
Langhoff and P.W. Langhoff, J. Chem. Phys. 76, 1043 (1982).

(e,2e) CHEMISTRY

K.T. Leung and C.E. Brion*

Department of Chemistry
The University of British Columbia
Vancouver, B.C., V6T 1Y6
Canada.

The notion that chemists might benefit by looking at molecular orbitals and bonding phenomena from the complementary momentum-space perspective was suggested by Coulson and Duncanson [1] some forty years ago. Momentum-space concepts were further studied by Epstein and Tanner [2]. The main driving force behind such a phenomenological development of momentum-space chemistry arose from the need to interpret Compton scattering data. The Compton profile $J(p)$ is related to an integral of the electron momentum density due to all the electrons in the target. (i.e. $J(p)=\int_p^\infty (\frac{1}{4\pi}\int \rho(\underline{p}')d\Omega)2\pi p'dp'$.) With the emergence of binary (e,2e) spectroscopy [3] in the last decade, experimental momentum densities of individual orbitals can now be measured and this has added a further need for such development of momentum-space descriptions of molecular electronic structure.

In a typical (e,2e) experiment, a gaseous target is ionized by high energy (>1200eV) electron impact and the scattered and ejected electrons are counted in coincidence under predetermined scattering kinematic conditions. The noncoplanar symmetric (e,2e) triple differential cross-section, in the plane wave impulse approximation and the target Hartree-Fock approximation [3], can be shown to be proportional to the orbital momentum density $\rho_j(p)$. i.e.

$$\frac{d\sigma}{dE_1 d\Omega_1 d\Omega_2} \propto \sigma_{Mott}\, S_j(\tfrac{1}{4\pi}\int \rho_j(\underline{p})d\Omega)$$

where σ_{Mott} is the Mott scattering cross-section and S_j is the spectroscopic factor. The momentum density is spherically averaged because of the random orientation of the gaseous targets used in the experiments. The momentum-space (p-space) wavefunction itself is, of course, related to the position-space (r-space) wavefunction by a Fourier transform. An (e,2e) experiment basically provides two fundamental pieces of information concerning the ionization process: namely, the binding energy spectrum and the spherically averaged momentum distribution. The binding energy spectrum has provided evidence of electron correlation and the breakdown of the independent particle ionization picture in atoms and molecules [3]. The momentum distribution (MD) has provided a means of identifying orbital symmetry (either s-type or p-type) associated with a particular binding energy peak. More importantly, the experimental MD provides a closer scrutinization of (and direct feedback to) literature SCF wavefunctions in the chemically important low momentum (i.e. large r) region.

We shall confine ourselves in the present article to a brief discussion of some of the interesting orbital MDs of a few small species (including noble gases [4], H_2 [5], CO_2, OCS and CS_2 [6]) determined recently using a high momentum resolution $(0.1a_0^{-1}$ fwhm) binary (e,2e) spectrometer [4]. In addition the relations between momentum-space and position-space densities are illustrated by an orbital density topographical approach [5,7] using density contour maps and 3-dimensional density surface plots [5]. Four general momentum-space relations [1,2, 5,7] are commonly used for guiding such an approach. These include: (i) inverse spatial reversal relation; (ii) symmetry invariance; (iii) molecular density directional reversal and (iv) molecular density oscillation.

* Research supported by the Natural Sciences and Engineering Research Council of Canada.

As an example of the usefulness of experimental MDs as a tool for testing the quality of theoretical wavefunctions, we show in fig. 1 the experimental MD for the H_2 $1\sigma_g$ orbital [5]. Comparison with theoretical MDs calculated using five different quality wavefunctions indicates that the variationally less superior wavefunctions, the double-zeta (DZ) and minimal basis set (MBS), are underestimating the small p region (i.e. MDs are too broad). Excellent agreement with Compton scattering data [8] is obtained. Moreover, fig. 1 shows the density contour maps and surface plots of the H_2 $1\sigma_g$ orbital calculated with the ext-HF wavefunction. The most striking feature is the change in the longitudinal orientation of the density lobe in p-space (the molecular density directional reversal) [5]. The 3-D surface plot also shows molecular density oscillations in p-space when a very low relative density value is used for the visualization (fig. 1 p-space 3-D plot, rhs). The density oscillations reflect nuclear geometry in p-space.

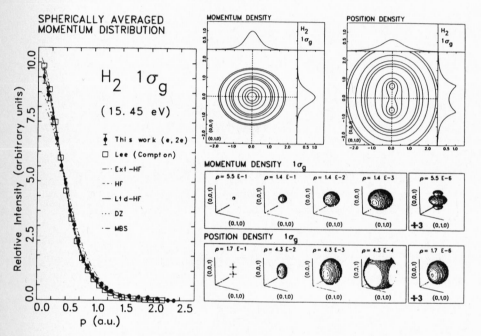

Figure 1

In the case of H_2, it is possible to derive the "experimental" spherically averaged momentum-space bond density [9] of the sigma bond from the normalized H_2 $1\sigma_g$ MD and the MD of the independent atom model. The wavefunction for the independent H atoms is the exact solution of the Schrodinger equation. Fig. 2 shows the derived experimental momentum-space bond density. A 3-dimensional visualization of momentum-space bond density using the HF wavefunction gives the "bone (negative density difference) in the donut (positive density difference)" picture. This drastically different view of the chemical bond in momentum-space is in marked contrast to the traditional "hamburger" model in position-space. The bonding picture in p-space complements the r-space bonding model and therefore provides a more complete understanding of chemical bonding phenomena since it emphasizes the chemically significant outer spatial regions (i.e. low momentum).

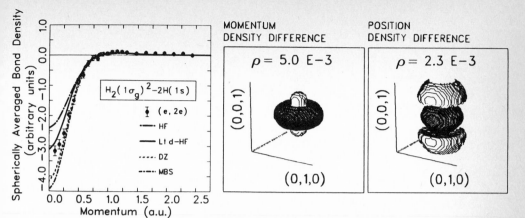

Figure 2

One of the main interests to chemists in the study of orbital MDs is the study of chemical trends in MDs of groups of similar orbitals in atoms and molecules. Fig. 3 shows the experimental MDs of the outermost np and ns electrons of the noble gases, along with the theoretical MDs computed using the corresponding Hartree-Fock quality wavefunctions. The observed MDs of both np and ns orbitals are narrower and are shifted to the low p region going down the group from Ne to Xe. This apparent decrease in average orbital momentum is of course complementary to the increase in the effective size of the orbital (in r-space) proceeding down the group.

SPHERICALLY AVERAGED MOMENTUM DISTRIBUTION

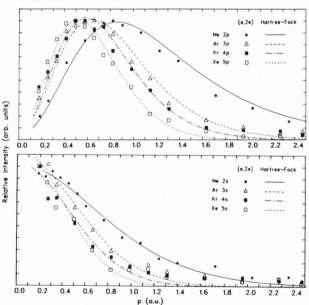

Figure 3

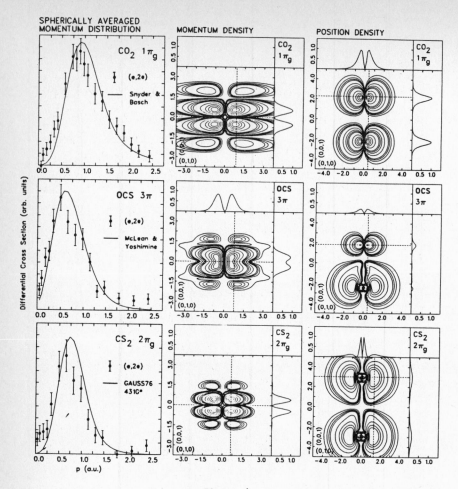

<u>Figure 4</u>

A further example is the valence isoelectronic triatomic series CO_2, OCS and CS_2 [6]. The MDs and their associated theoretical density maps of the first and fourth outermost occupied orbitals of the group are shown in figs. 4 and 5 respectively. In fig. 4 the p-space density maps clearly show the effect of replacing O with S. The inverse spatial reversal property is clearly illustrated for the essentially nonbonding π_g orbitals of CO_2 and CS_2. The expansion of the r-space density in going from O2p in CO_2 to S3p in CS_2 is associated with the contraction in the p-space densities. This p-space density contraction is clearly observed in the experimental MDs, which show a shift of maximum from $\simeq 0.9a_0^{-1}$ in CO_2 to $\simeq 0.7a_0^{-1}$ in CS_2 as well as a general narrowing in the CS_2 MD. This shift of maximum and general narrowing of the MDs in going from CO_2 to CS_2 are qualitatively similar to the case of atoms (c.f. Ne2p and Ar3p in fig. 3). Similar observations can also be made in the case of the CO_2 $4\sigma_g$ and CS_2 $6\sigma_g$ orbitals (fig. 5). It is interesting to note that the shapes of these MDs clearly reflect the s-p composite nature of these orbitals. The second relative maximum observed in the MD is essentially due to the

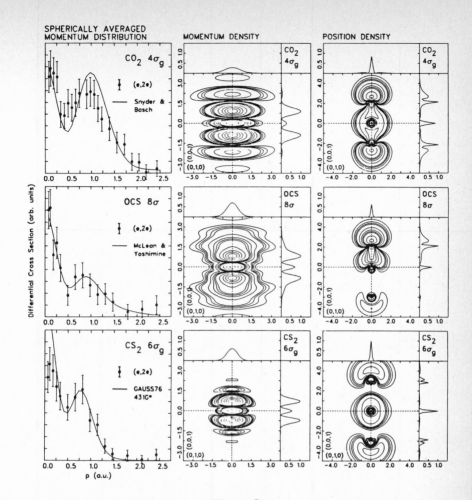

SPHERICALLY AVERAGED MOMENTUM DISTRIBUTION

MOMENTUM DENSITY

POSITION DENSITY

Differential Cross Section (arb. units)

CO₂ 4σg

(e,2e)

Snyder & Bosch

OCS 8σ

(e,2e)

McLean & Yoshimine

CS₂ 6σg

(e,2e)

GAUSS76 431G*

p (a.u.)

Figure 5

O2p in CO_2 (and S3p in CS_2.) One important difference between OCS and its valence isoelectronic counterpart is the relative s to p composition as indicated approx- imately by the ratio of the maxima at p=0 (s-type) and at p>0 (p-type). Also of interest is the general deviation of the appearance of the p-space density maps of the 3π and 8σ of OCS from those of the corresponding orbitals of CO_2 and CS_2.

References

[1]   C.A. Coulson and W.E. Duncanson, Proc. Cambridge Phil. Soc., 1941, 37, 55,67, 74,397,406; 1942, 38, 100; 1943, 39, 180.
[2]   I.R. Epstein and A.C. Tanner, in "Compton Scattering", B.G. Williams (ed.), McGraw-Hill, New York, 1977, pp209.
[3]   I.E. McCarthy and E. Weigold, Phys. Rep. 27C, 1976, 276.
[4]   K.T. Leung and C.E. Brion, in press (Chem. Phys.) 1983.
[5]   K.T. Leung and C.E. Brion, in press (Chem. Phys.) 1983.
[6]   K.T. Leung and C.E. Brion, to be published.
[7]   J.P.D. Cook and C.E. Brion, in "Momentum Wave Function-1982", E. Weigold (ed.), AIP Conf. Proc. No. 86, AIP, New York, 1982, pp278.
[8]   J.S. Lee, J. Chem. Phys., 1977, 66, 4906.
[9]   K.T. Leung and C.E. Brion, submitted to JACS, 1983.

FOURIER TRANSFORM OF SPHERICALLY AVERAGED
MOMENTUM DENSITIES*

M. A. Coplan and D. J. Chornay
Institute for Physical Science and Technology

J. H. Moore and J. A. Tossell
Department of Chemistry

N. S. Chant
Department of Physics

University of Maryland, College Park, MD   20742

The first (e,2e) experiments had as their objective the demonstration of the feasibility of the technique.[1] Energy resolution was several eV and the signal rate was low, however, the early measurements clearly showed that the electron knock-out coincidence technique could be used for obtaining single electron momentum densities. Once the feasibility of the technique was established new instruments were constructed with better energy resolution and greater efficiency.[2,3,4,5] For structure measurements the noncoplanar symmetric geometry was preferred because of the ease with which the data could be converted to momentum densities within the context of the plane wave impulse approximation. During this phase a great deal of data was accumulated on atoms and small molecules. The emphasis was on the assignment of electronic states, the investigation of satellite structure and the comparison between momentum densities calculated from atomic and molecular wavefunctions and the corresponding experimental measurements.[6] In this second phase there was a dramatic increase in the quality of the experimental data, however, the data were not of sufficient precision to make more than qualitative comparisons with theory. An important criticism of the technique was that all the experimental data looked qualitatively similar and most analyses were essentially descriptive.

Compared to photoelectron spectroscopy where one routinely obtains only two parameters for each bound electron – the binding energy and the relative ionization efficiency – (e,2e) spectroscopy adds another dimension, the single electron momentum density. The challenge is to obtain the densities with sufficient precision to permit quantitative comparisons with theory.

At Maryland we use a multiple detector spectrometer[7,8] to obtain a 25-fold increase in collection efficiency over the conventional two detector configuration, while at the same time eliminating systematic errors due to changes in electron currents and gas pressure during the course of an experiment. The magnetic field has been reduced to less than 0.2 milligauss over the entire volume of the

*Research supported by NSF grant CHE 8205884 and the Computer Science Center of the University of Maryland.

spectrometer with a 3-axis helmholtz coil and a double magnetic shield. The gains of each of the detectors are separately adjustable and are continuously measured while data reduction is done in real time so that the progress of an experiement can be monitored.

As an example of the quality of the data, Figure 1 shows the momentum distribution for the t_{2g} orbital in $Cr(CO)_6$. This orbital is predominately Cr 3d. These

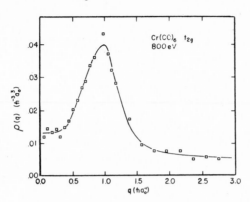

Figure 1. Momentum density for the t_{2g} orbital of $Cr(CO)_6$ at 800 eV incident energy.

data were obtained in three days and are comparable in quality to those obtained for H_2 just a few years ago. This is all the more remarkable considering the low vapor pressure of the compound and the large number of valence electrons which contribute to the background noise.

With better data we are in a position to look more carefully at interpretation. We are pursuing a scheme based on the Fourier transform of the momentum density.[9,10] The transform function $B(r)$ is also the autocorrelation function of the spatial wavefunction and can be calculated directly from theoretical position space wavefunctions. An attractive feature of $B(r)$ is that is can be interpreted in terms of the familiar concepts of wavefunction amplitudes in position space. There are, however, two difficulties associated with this approach - one experimental and one theoretical: the Fourier transformation of the experimental data requires that the momentum distribution be weighted by the square of the momentum so that the large momentum components of the experimental distribution play an important role in the determination of $B(r)$. Unfortunately, the large momentum values of the distribution are relatively less well determined than the low and intermediate values and there is evidence that the approximation which permits one to equate the experimental measurements with momentum densities - the plane wave approximation - is in error at high momenta. The theoretical difficulty is of a philosophical nature; to interpret experimentally determined $B(r)$ functions wavefunction amplitudes are needed, but the

purpose of obtaining $B(r)$ is to get wavefunction information. We have taken a
pragmatic approach in this case. An approximate wavefunction is used to calculate
a theoretical $B(r)$, and the difference between the experimental and theoretical
functions are interpreted in terms of the initial approximate wavefunction. In some
cases it is possible to go further when investigating the difference between two
orbitals in a single molecule or analogous orbitals in two different molecules. As
an example, Figures 2 and 3 show the measured momentum distributions for the lone

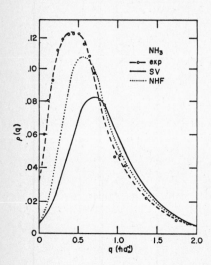
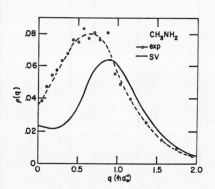

Figure 2. Momentum density for
the lone pair orbital of ammonia.
The experimental data are shown
along with split valence (SV)
and near Hartree-Fock (NHF) calculations.

Figure 3. Momentum densities
for the lone pair orbital of
methylamine (CH_3NH_2). The experimental data and a split
valence (SV) calculation are
shown.

pair electrons of ammonia and methylamine. Also shown in the figures are calculations
based on split valence (SV) and near Hartree-Fock (NHF) wavefunctions. It is well
known that the basicity of the two moleucles depends primarily on the lone pair
electrons and that methylamine is more basic than ammonia. Calculations[11] have
attributed this difference to a significant contribution from the methyl group to the
major lobe of the lone pair in methylamine. Figure 4 shows experimental and theoretical difference functions $\Delta B(r)$ for the corresponding lone pair orbitals in ammonia
and methylamine. Keeping in mind that similarities between the orbitals will tend
to cancel, the two functions are qualitatively very much alike. Since the wavefunctions from which the calculated $\Delta B(r)$ reproduce the H2s and C2p contributions to
the N2p component of the lone pair of methylamine, we feel that the agreement between
experiment and theory is good evidence for the validity of the method.

While the use of the difference function $\Delta B(r)$ minimizes inaccuracies in $B(r)$
due to errors in the high momentum components, it is important to try to evaluate
the magnitude of the errors and correct them if possible.

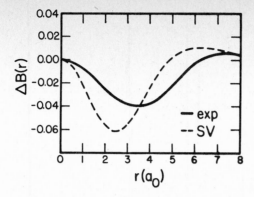

Figure 4. B(r) difference function ΔB(r) for the lone pair
orbitals of ammonia and methylamine. The experimental results
are compared with a split valence
(SV) calculation.

In Figure 5 the measured momentum density for the 1s electrons of helium is compared with a high quality CI wavefunction calculated by Smith and Mrozek.[12] On the

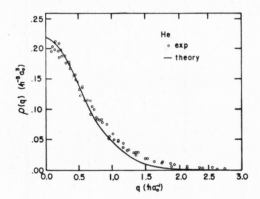

Figure 5. Momentum density for the
1s orbital helium. The data were obtained at incident energies of 400,
800 and 1200 eV. The theortical
curve is based on a CI wavefunction.

scale of the figure, agreement appears to be excellent. The picture changes, however,
when $\rho(q)q^2$ is plotted against q as shown in Figure 6. Weighting the momentum density by q^2 gives a more realistic picture of the relative volume in momentum space
occupied by the wavefunction. Also shown on the figure is the result of correcting
the experimental measurements for distortions in the incoming and outgoing electron
waves.[13] Noteworthy is the fact that the corrected experimental results are smooth

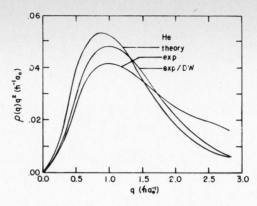

Figure 6. Weighted momentum density
for the 1s orbital of helium. Theoretical results based on a CI wavefunction
are compared with experiment and exper-
ment corrected using a distorted wave
(DW) calculation.

at high momenta while the uncorrected data undulate, possibly due to angular momentum
components of the incident and outgoing waves not properly accounted for in the plane
wave approximation. We are optimistic that this procedure can be generalized to permit accurate $B(r)$ functions to be obtained from the experimental data. To do this
one needs for the distorted wave calculation the incident electron-target distorting
potential as well as the overlap between the initial and final states of the system.
For helium these were easily obtained, but for more complex systems it will be
necessary to use a combination of approximate wavefunctions and empirical data obtained from subsidiary elastic scattering experiments now under way in our laboratory.

The course of the (e,2e) experiment has followed a familiar route; as the
quality of experimental measurements has increased, new questions have been raised
which have required theoretical and computational advances which in turn have placed
new demands on the experiments and opened the way for new ones.

REFERENCES

1. U. Amaldi, A. Egidi, R. Marconero and G. Pizzela, Rev. Sci. Instrum. 40, 1001 (1969).

2. R. Camilloni, A. Giardini-Guidoni, R. Tiribelli and G. Stefani, Phys. Rev. Lett. 29, 618 (1972).

3. E. Weigold, S. T. Hood and P. J. O. Teubner, Phys. Rev. Lett. 30, 475 (1973).

4. M. A. Coplan, J. H. Moore and J. A. Tossell, J. Chem. Phys. 68, 329 (1978).

5. S. T. Hood, A. Hamnett and C. E. Brion, Chem. Phys. Lett. 29, 252 (1976).

6. J. H. Moore, J. A. Tossell and M. A. Coplan, Acc. Chem. Res. 15, 192 (1982); I. E. McCarthy and W. Weigold, Phys. Rep. 27, 275 (1976).

7. J. H. Moore, M. A. Coplan, T. L. Skillman and E. D. Brooks, Rev. Sci. Instrum. 49, 463 (1978).

8. T. L. Skillman, E. D. Brooks, M. A. Coplan and J. H. Moore, Nucl. Instrum. Methods 159, 267 (1978).

9. W. Weyrich, P. Pattison and B. G. Williams, Chem. Phys. 41, 271 (1979).

10. J. A. Tossell, J. H. Moore and M. A. Coplan, Elect. Spect. Rel. Phenom. 22, 61 (1981)

11. W. G. Henderson, J. L. Beauchamp. D. Holtz and R. W. Taft, J. Amer. Chem. Soc. 94, 4724 (1972).

12. V. H. Smith, Jr. and J. Mrozek, private communication.

13. N. S. Chant and P. G. Roos, Phys. Rev. C, 15, 57 (1977).

HIGH RESOLUTION STUDIES OF INNER-SHELL EXCITED STATES

OF ATOMS AND MOLECULES BY ELECTRON IMPACT EXCITATION

George C King

Department of Physics, Schuster Laboratory,
Manchester University, Manchester, UK.

1.   Introduction

Inner-shell excited states of atoms and molecules are formed when an inner-shell electron is promoted to an unoccupied valence or Rydberg orbital, for example the excitation of a 2p electron in Ar or a 1s atomic electron in N_2. Traditionally these states have been studied by photoabsorption measurements using energetic photons (for example [1,2]). These inner-shell transitions can also, however, be studied by electron energy-loss spectroscopy and in fact this technique can have important advantages over photoabsorption measurements. Van der Wiel et al [3] employed the electron energy-loss technique and observed inner-shell transitions in N_2 and CO. This work was followed by a systematic and comprehensive study of inner-shell transitions in molecules by Brion and collaborators using the electron energy-loss technique at an energy resolution of approximately 0.5eV (for a recent review see [4]). Subsequently a significant improvement in resolution was obtained by the Manchester group (for example [5,6]) who obtained a resolution of 0.07eV in electron energy-loss measurements in a number of atoms and molecules. These latter studies illustrated one of the important advantages of the technique, namely its superior resolution to photo-absorption measurements for state excitation energies of above about 200eV. This allowed a full investigation of the parameters of inner-shell states including their energies, natural widths, and in the case of molecules, their vibrational spacings and equilibrium internuclear separations.

A second and perhaps more important advantage of electron impact excitation is its ability to induce electric-dipole-forbidden transitions. This can occur when the value of the incident electron energy is reduced to a value close to the excitation energy of the state, when optical selection rules are considerably relaxed. Recently the electron energy-loss technique has been applied for the first time to the study of these electric-dipole-forbidden transitions [7]. In that work an inner-shell spin-forbidden transition in N_2 and a parity-forbidden transition in Ar were studied. These transitions were observed by using incident electron energies as low as 1.15 times the excitation energy of the inner-shell states. Recent progress in high resolution studies of inner-shell states in atoms and molecules will be discussed with particular reference to the observation of electric-dipole-forbidden transitions.

2.   The technique of electron energy-loss spectroscopy

In this technique an electron beam is produced and energy selected to provide a well defined beam with an energy spread of typically 0.05eV and with an adjustable mean energy. The ability to change the energy of the incident electron is very powerful and is a degree of freedom not available in photoabsorption measurements. The electron beam is passed through a gas cell containing the target atoms or molecules and those electrons that are inelastically scattered within a small angular range in the forward direction are analysed in energy, again with a resolution of typically 0.05eV. The spectrum of electron energy loss obtained corresponds directly to the energy levels of the target gas.

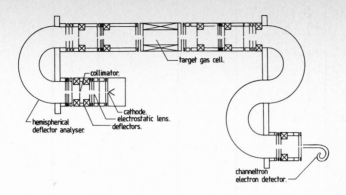

FIGURE 1:   Schematic diagram of a high resolution electron energy-loss spectrometer.

The apparatus used for the studies is shown in Figure 1. It consists of an electron spectrometer and utilises 180° hemispherical electrostatic deflectors of mean radius 2.54cm to select and analyse the electron energy. Of note is the use of two hemispherical deflectors placed in tandem in the electron energy analyser. This arrangement provides very efficient rejection of unwanted scattered electrons which is an important consideration since the cross sections for exciting inner-shell electrons are low and therefore the detected signals are small. Combinations of triple-aperture electrostatic lenses are used to transport and focus the electron beam from the electron source, a directly heated tungsten filament, to the energy selector and from the selector to the target region and similarly through the analyser section of the spectrometer. The lenses also enable the incident energy of the electrons at the target to be varied over the range 50 to 1500eV. In all of the results discussed the scattering angle of the detected electrons is close to 0°.

The inner-shell states that can be studied most usefully with an energy resolution of 0.05eV are those having natural decay widths, Γ, that are in the range from approximately 0.05 to 0.2eV. Many states having a vacancy in the sub-shell immediately below the valence sub-shell (for example a vacancy in the 3s sub-shell of Ar) have values of Γ in this range as do most states having vacancies in the next lower lying sub-shell or full shell (for example a 2p vacancy in Ar). Their excitation energies are typically in the range from 50 to 500eV. States with deeper lying vacancies tend to have natural widths greater than 0.2eV and are therefore better studied with lower-resolution techniques.

3.   Comparison between electron impact and photoabsorption measurements

At high values of incident electron energy, electron impact excitation is analogous to photoabsorption. This analogy was first established by Bethe [8] who showed that for fast electrons there is a quantitative relationship between the differential electron scattering cross section and the generalised oscillator strength, $f_n(K)$ (see also [9]):

$$\frac{d\sigma}{d\Omega} = \frac{4k_n \, R \, f_n(K)}{E_n \, k_o \, K^2} \tag{1}$$

Here $\hbar k_o$ and $\hbar k_n$ are the momenta of the incident and scattered electrons respectively, $\hbar K$ is the momentum transferred to the target atom or molecule, R is the Rydberg energy and E_n is the excitation energy. When K is small compared with a_o^{-1} (where a_o is the Bohr radius) the generalised oscillator strength can be expanded in powers of K (see also [10]):

$$f(K) = f^{opt} + Kf^{(1)} + K^2 f^{(2)} + \ldots \tag{2}$$

where f^{opt} is the familiar optical oscillator strength. The second and higher terms of the expansion become negligible when E_n is much smaller than the incident electron energy and when the angle of scattering is small. Then the electron scattering cross section is directly related to the optical oscillator strength, and an electron energy-loss spectrum effectively simulates the photoabsorption spectrum. In inner-shell excitation experiments E_n is usually large enough that linear and quadratic terms in equation (2) cannot be ignored, and then the electron energy loss spectrum contains peaks due to non-electric-dipole transitions. Such transitions were reported by King et al [6] who observed electric-quadrupole transitions in Ar, Kr and Xe. The effect can be enhanced by a judicious use of incident electron energy and scattering angle. Thus electron impact excitation can excite transitions that cannot be seen with photoabsorption.

A further process that is possible with electron impact excitation and not photo-absorption is electron exchange. Thus as the incident electron energy approaches the excitation energy of the state it becomes possible for the incident electron and the target electron to exchange places. Singlet-to-triplet transitions which are of course also electric-dipole-forbidden, can then be excited.

4.   Inner-shell excitation in molecules

Examples of inner-shell energy-loss spectra, obtained in N_2, are shown in Figure 2.[7]. These spectra correspond to the promotion of a 1s electron from one of the nitrogen atoms to the lowest unoccupied $2p\pi$ orbital of the molecule. The top spectrum corresponds to the excitation of the singlet state $(1s)^{-1}(2p\pi)^1\Pi$. This state has been studied with lower resolution by Wight et al [11] who observed that this peak has an intensity which is much greater (by a factor of the order of 10) than the sum of the intensities of all the other discrete peaks below the ionization threshold. This is attributable [12] to the fact that the $2p\pi$ orbital is highly localised, being confined to the vicinity of the molecular core by a centrifugal barrier in its d-wave component.

The existence of this barrier is also responsible for the well known shape resonance in N_2 at 2.3eV [13]. At energies above about 1000eV the $(1s)^{-1}(2p\pi)^1\Pi$ is the only structure observed in this region. As the incident energy is reduced, however, a low energy structure becomes evident and at the lowest values of incident energy used it becomes dominant. This structure corresponds to the promotion of a 1s nitrogen electron to the $2p\pi$ orbital to give the $(1s)^{-1}(2p\pi)^3\Pi$ state of N_2:

$$e_{inc} + N_2 \qquad N_2(1s)^{-1}(2p\pi)^3\Pi + e_{scatt} \tag{3}$$
$$\uparrow \quad \uparrow\downarrow \qquad \uparrow\uparrow \qquad\qquad\qquad \downarrow$$

The arrows indicate the spin states of the incident and 1s target electrons. Such transitions can only be excited by near-threshold electron impact and this study represents the first observation of this triplet state in N_2. Of note is the resolution of these spectra, namely 0.065eV. This corresponds to a resolution of 0.0005nm at a wavelength of 31nm which is still far from obtainable with present photon sources. These spectra and corresponding spectra in other molecules (see for

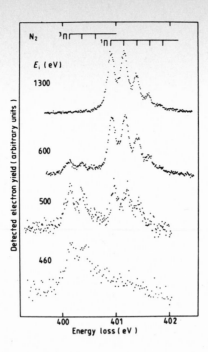

FIGURE 2: Energy-loss spectra of N_2 obtained at the indicated values of incident electron energy and with a resolution of 65meV. The lower- and higher-energy vibrational structures correspond to excitation of the $(1s)^{-1}(2p\pi)^3\Pi$ and $(1s)^{-1}(2p\pi)^1\Pi$ states of N_2 respectively. From Shaw et al [7].

example, [7,14]) give a variety of new and important information, as outlined below.

The spectra of figure 2 can be analysed to yield the molecular parameters ω_e, $\omega_e x_e$, and R_e and also the natural line width Γ of the molecular states. Table 1 shows a comparison of these parameters for the singlet and triplet states, where it may be seen that these parameters are similar for both states although the value of R_e is slightly smaller for the $^3\Pi$ state than for the $^1\Pi$ state.

State	ω_e (meV)	$\omega_e x_e$ (meV)	R_e (Å)	Γ (meV)
$(1s)^{-1}(2p\pi)^3\Pi$	242(4)	1.8	1.155(6)	123(10)
$(1s)^{-1}(2p\pi)^1\Pi$	236(2)	1.8	1.162(3)	123(10)

TABLE 1: Derived properties of two inner-shell excited states of N_2. The numbers in brackets indicate the standard error in the last significant figures. From Shaw et al [7].

The spectra also give a direct and accurate determination of the singlet-triplet splitting which is found to be 0.82 $\pm$ 0.01eV. Such a measurement is useful because it gives the magnitude of the exchange interaction between the excited electron and the inner-shell hole. In fact the singlet-triplet splitting is equal to twice the exchange integral K. Rescigno [15] has calculated the singlet-triplet splitting in N_2 to be 0.9eV. There have been experimental determinations of singlet-triplet splittings of inner-shell valence states in other molecules and these are shown in Table 2 together with calculated values [14]. It is interesting to note that the

	Experiment (eV)	Theory (eV)
N_2^{K*}	0.82 $\pm$ 0.01 [a]	0.9 [b]
$C^{K*}O$	1.46 $\pm$ 0.01 [c]	2.20 [d]
$C^{K*}O_2$	1.2 [c]	1.15 [e]
$N^{K*}NO$	small [c]	small [e]
$NN^{K*}O$	0.4 [c]	

a Shaw et al [7], b Rescigno [15], c Shaw et al [14],
d Kondratenko et al [24], e Schwarz and Buenker [21].

TABLE 2: Experimental determinations and theoretical calculations of inner-shell singlet-triplet splittings.

singlet-triplet splittings in N_2 and CO are quite different although the two molecules are isoelectronic and have the same equivalent core molecule (see later discussion). Presumably this occurs because the 1s hole is more tightly bound around the nitrogen atom than it is around the carbon atom. Consequently the overlap between the hole and the excited $2p\pi$ orbital, the latter being the same in both cases, will differ, resulting in a smaller exchange interaction in N_2. N_2O provides another interesting case, since the two nitrogen nuclei are non-equivalent in the NNO structure. The excited valence orbital has a larger overlap with the central nitrogen nucleus than the terminal one and so the exchange interaction and hence singlet-tripet splitting would be expected to be greater for the central nitrogen nucleus, as observed experimentally.

There has been much discussion (see, for example, [16,17]) about whether the hole is localised at the site of one of the two nitrogen nuclei. If the hole is delocalised then u-g symmetry is appropriate and a u-g splitting would be expected. Rescigno and Orel [16] for example, have calculated this splitting to be 0.06eV. The $\sigma_u 1s \rightarrow \pi_g 2p$ transition would be electric-dipole allowed while the $\sigma_g 1s \rightarrow \pi_g 2p$ transition would be electric-dipole forbidden. In earlier work on the $(1s)^{-1}(2p\pi)^1\Pi$ state of N_2, [17], King et al did not see any evidence for the occurrence of two closely lying states although the predicted splitting is smaller than the natural widths of the states and the finite resolution of the measurements. In the spectra of figure 2, however, larger values of momentum transfer are achieved, increasing from 1.65au at 1300eV incident energy to 3.73au at 460eV which might lead to an enhancement of electric-dipole-forbidden transitions (as predicted by Rescigno and Orel). The enhancement might lead, for example, to a change in the areas of the vibrational levels, or to changes in the peak energies or widths. However, within the statistical accuracy of the measurements no changes of this type are observed suggesting that the u-g splitting is small. Other authors have studied the $(1s)^{-1}(2p\pi)^1\Pi$ state of N_2

using large values of momentum transfer, for example [18] but they did not have the resolution to see the u-g splitting.

The inner-shell spectra also give a better interpretation of the equivalent core model (see for example [1,2,19]). In this model the energy levels of a molecule with an inner-shell hole would be similar to those of a molecule with a fully occupied electronic core but with a nuclear charge increased by one unit. For example, in the case of inner-shell excitation in N_2 with the 1s hole localised on one of the nitrogen nuclei, the outer electrons of the molecule move in a mean electrostatic potential similar to that of the outer electrons of the NO molecule which is thus the "equivalent core molecule". This model has been successfully applied to a number of molecules. For example, inner-shell spectra obtained in Cl_2 have been used to obtain information about its equivalent core molecule ArCl which because of its unbound ground state cannot be studied directly [20]. Although the model has been successful with Rydberg states of the molecules there have been discrepancies with the lowest valence states (see for example [11,21]). This has occurred partly because exchange interactions have often been neglected usually because their magnitudes have not been known. For example, N_2 inner-shell states and NO outer-shell states are not completely equivalent because the excited electron in NO does not have the inner-shell hole to give rise to a large singlet-triplet splitting as the N_2 molecule has. (The Rydberg states in N_2 do not have an appreciable overlap with the 1s hole and so any singlet-triplet splitting is small). The energies of the singlet and triplet $(1s)^{-1}(2p\pi)$ states of N_2 and CO can be used to calculate the weighted mean energy of the $(1s)^{-1}(2p\pi)$ configuration of each molecule, and hence to calculate the mean binding energy of this configuration with respect to the ion core configuration $(1s)^{-1}$. This mean binding energy is found to be the same, 9.69eV, for both molecules. This appears to indicate that the ion cores N_2K^* and $C^{K^*}O$ provide the same environment for the $2p\pi$ electron, as they would in the equivalent-core model since they are both equivalent to NO^+.

A further interesting aspect of the inner-shell spectra of figure 2 is their relationship to the inner-shell resonances observed in N_2 by King et al [22]. These resonances consist of an extra electron bound to the inner-shell excited states, and it was originally assumed that the appropriate parent state was the inner-shell singlet state, the only one observed at that time. A table of such binding energies obtained for a range of molecules [23] is shown in Table 3. It may be seen that there is a fairly large spread in the magnitude and sign of these binding energies. The energy loss

	Energy	Binding energy (a)	Binding energy (b)
$N_2^{K^*}$	400.4	0.5	−0.3
$N^{K^*}NO$	401.2	−0.8	−0.8
$NN^{K^*}O$	403.2	0.6	−0.2
$C^{K^*}O$	286.1	1.3	−0.1
$C^{K^*}O_2$	289.4	0.4	−0.8

(a) Measured with respect to inner-shell singlet state.
(b) Measured with respect to inner-shell triplet state.

TABLE 3: Energies and binding energies (eV) of inner-shell resonances.

spectra of figure 2 show that the intensity of the triplet state relative to the singlet state increases dramatically as the incident electron energy approaches threshold indicating that the near-threshold cross section for the triplet state is much larger than that of the singlet state. If it can be considered that the resonance consists of an extra electron attached to a parent state then it might reasonably be expected that the resonance is associated with the triplet state. If the binding energy of the resonances are measured with respect to the triplet state then a more consistent set of values is obtained as shown in Table 3. On the other hand the resonance can be interpreted in terms of the grandparent model, so that the two outer electrons couple together to form a correlated pair that couples to the N_2^{K+} core. The energy with which the $(2p\pi)^2$ pair are bound to the ion core is then 9.3eV, which is close to the mean binding energy of a single $(2p\pi)$ electron to the core, namely 9.69eV, as deduced above.

5.   Inner-shell excitation in atoms

High resolution energy-loss spectra corresponding to inner-shell excited state of the rare gases Ar, Kr and Xe have been reported by King et al [6]. In these spectra evidence for the excitation of electric-dipole forbidden transitions was observed, such as the 2p→4p transition in Ar. More recently Shaw et al [7] investigated electric-dipole forbidden transitions in Ar also looking in electron energy-loss but using low values of incident electron energy. Their results in Ar are shown in figure 3.   The peaks in the spectra correspond to the promotion of a $2p_{3/2}$ or $2p_{1/2}$

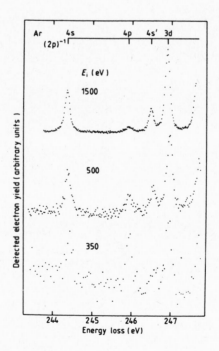

FIGURE 3:   Energy-loss spectra of argon obtained at the indicated values of incident electron energy, and with a resolution of 65meV. From Shaw et al [7].

electron to unoccupied Rydberg orbitals. The features observed in the 1500eV
incident energy spectrum are similar to those described by King et al [6]. The
major point of note is the dramatic rise in the relative intensity of the electric-
dipole forbidden 2p→4p transition as the incident electron energy is reduced. The
observed peak will correspond to several different J values that will be separated
by spin-orbit and electrostatic terms but Shaw et al [7] estimated that these splittings are small compared to the natural line widths of the states and the finite
resolution of the measurements. The ratio of the intensities of the $2p_{3/2}$→4p and
$2p_{3/2}$→4s transitions is plotted in figure 4 as a function of the dimensionless
quantity Ka_0 where $\hbar K$ is the momentum transfer. It may be seen that the intensity

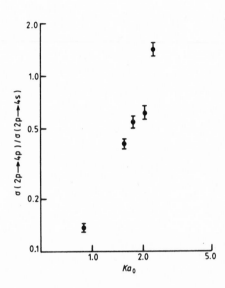

FIGURE 4: The dependence of the ratio of intensities of the observed peaks corresponding to the transitions 2p→4p (parity forbidden) and 2p→4s (electric-dipole
allowed) on the dimensionless quantity Ka_0, where $\hbar K$ is the momentum transfer and a_0
is the Bohr radius. From Shaw et al [7].

ratio is approximately proportional to K^2. From the Born approximation which gives
the leading power of K in an expansion of the differential cross section it may be
found that the J=0 and J=2 levels of the $(2p_{3/2})^{-1}4p$ configuration have differential
cross sections proportional to K^0 corresponding to electric-quadrupole transitions.
Further, the ratio of this cross section to that of the $(2p_{3/2})^{-1}4s$ level should be
proportional to K^2 at sufficiently high energies. This is in fact approximately
the observed dependence although the Born approximation may apply only poorly at the
incident energies used in the work of Shaw et al. Most recently Shaw et al [14]
have investigated inner-shell transitions near the M-edge of Kr and have been able
to resolve individual J levels and their relative differential cross sections as a
function of incident electron energy.

6. Conclusion

Electron impact excitation has considerable advantages in the study of inner-shell excitation in atoms and molecules. High resolution in the measurements has been realised enabling new inner-shell states and fine structure to be resolved. Now the ability of electron impact excitation to excite dipole-forbidden states is being exploited and this is resulting in a wealth of new information about previously unobserved states.

7. References

1. M J-A Prins, Physica 1 (1934) 1174.

2. M Nakamura et al, Phys.Rev.A178 (1969) 80.

3. M J Van der Wiel, Th M El Sherbini and C E Brion, Chem.Phys.Letts. 7 (1970) 161.

4. C E Brion "Physics of Electronic and Atomic Collisions", ed S Datz, North Holland Publishing Company (1982) 579.

5. M Tronc, G C King, R C Bradford and F H Read, J.Phys.B: Atom.Molec.Phys. 9 (1976) L555.

6. G C King, M Tronc, F H Read and R C Bradford, J.Phys.B: Atom.Molec.Phys. 10 (1977) 2479.

7. D A Shaw, G C King, F H Read and D Cvejanović, J.Phys.B: Atom.Molec.Phys. 15 (1982) 1785.

8. H Bethe, Ann.Phys. (Leipzig) 5(5) (1930) 325.

9. M Inokuti, Rev.Mod.Phys. 43 (1971) 297.

10. W M Huo, J.Chem.Phys. 71 (1979) 1593.

11. G R Wight, C E Brion and M J Van der Wiel, J.Electron.Spectros. and Rel.Phen. 1 (1972/3) 457.

12. J L Dehmer and D Dill, J.Chem.Phys. 70 (1976) 3390.

13. G J Schulz, Rev.Mod.Phys. 45 (1973) 378.

14. D A Shaw, G C King and F H Read, J.Phys.B: Atom.Molec.Phys. (1984) to be published.

15. T N Rescigno, Private Communication (1983) and this volume.

16. T N Rescigno and A E Orel, J.Chem.Phys. 70 (1979) 3390.

17. G C King, F H Read and M Tronc, Chem.Phys.Letts. 52 (1977) 40.

18. R Fantoni and G Stefani, 4th General Conference of the European Physical Society, Trends in Physics, York, England (1978).

19. W H E Schwarz, Angew.Chem.Int.Edn. 13 (1974) 456.

20. D A Shaw, G C King and F H Read, J.Phys.B: Atom.Molec.Phys. 13 (1980) L723.

...

21. W H E Schwarz and R J Buenker, Chem.Phys. 13 (1976) 153.

22. G C King, J W McConkey and F H Read, J.Phys.B: Atom.Molec.Phys. 10 (1977) L541.

23. G C King, J W McConkey, F H Read and B Dobson, J.Phys.B: Atom.Molec.Phys. 13 (1980) 4315.

24. A V Kondratenko, L N Mazalov, F Kh Gel'mukhanov, V I Audeev and E A Saprykhina, J.Struc.Chem. 18 (1977) 437 (Zh.Strukt.Khim. 18 (1977) 546).

K-EDGE EXCITATION IN N_2, NO, N_2O and BF_3 STUDIED BY ANGULAR RESOLVED ELECTRON ENERGY-LOSS SPECTROSCOPY

R. Camilloni, E. Fainelli, G. Petrocelli* and G. Stefani

F. Maracci and R. Platania

Istituto Metodologie Avanzate Inorganiche C.N.R.

Area Ricerca di Roma, CP 10, 00016 Monterotondo, Italy

INTRODUCTION

Molecular inner-shell excited states have been largely investigated by photoabsorption and electron energy-loss experiments. For both cases the molecular absorption spectrum in the vicinity of an inner-shell absorption edge shows large variations with respect to the correspondent atomic spectrum.

The most relevant differences are:

i) A large fraction of the transition intensity "below threshold" is concentrated in a single, very intense transition rather then in a regular Rydberg series.

ii) Just "above treshold" intense transitions a few eV wide are overimposed on the smooth continuum.

Transitions of this kind are due to the anisotropy of the molecular potential, which results in an effective barrier to the motion of the slow escaping electron. Both a centrifugal component in the electron-molecule scattering Hamiltonian or different electronegativity of the atoms in a molecule can create such a barrier. The effect is to localize few molecular orbitals close to the ionisation continuum, where the slow electron can be trapped. A review of the electron impact inner-shell excited states spectroscopy is given by King (1), in his contribution to this book, and to that paper we shall refer the reader for all the generalities concerning the subject. In the following we shall report experimental results recently obtained at our Institute on the $1sN \rightarrow n\pi^*$ transitions in N_2, NO and N_2O and on BK* excited states in BF_3.

Inner-shell excited states for these molecules have been already investigated either by photoabsorption or by small momentum transfer energy-loss spectroscopy. Transition energies, vibrational structures and, in some case, the Oscillator Strength for these transitions are known. In the present work they are investigated by momentum-transfer resolved energy-loss spectroscopy. Through the angular dependence of the cross-section, the Generalized Oscillator Strength (GOS) has been measured over a wide range of momenta with the following aims:

i) to determine the optical limit of the GOS by suitable extrapolation to zero momentum transfer of its angular dependence;

ii) to clarify the correctness of the localized model for K-shell holes.

iii) to investigate the relevance of the "chemical environment", which is known to affect transition energies, to the angular behaviour of the GOS.

EXPERIMENTAL

Within the framework of the First Born Approximation (2), the target response to the electron impact excitation is described in terms of a bidimensional GOS $f(\overline{K}, \varepsilon)$, where $\overline{K}$ is the momentum transfer and ε is the energy lost by the incident electron. The equivalent quantity in photoabsorption is the Optical Oscillator Strength (OOS) $f_0(\omega)$, to which the GOS converges in the limit of vanishing momentum transfer. On this ground optically related quantities can be extracted from electron energy-loss experiments and compared with photoabsorption experiments results. Conversely, in the region of momentum transfer different from zero, where also optically forbidden transitions are excited, unique informations can be derived from the momentum transfer dependence of the GOS. This quantity is related to the differential energy-loss cross section by the formula:

$$\frac{d^2\sigma}{d\Omega\, d\varepsilon} = 2 \frac{k_1}{k_0} \frac{1}{K^2} \frac{1}{\varepsilon} f(K, \varepsilon) \qquad (1)$$

k_0 and k_1 are the initial and final momentum of the incident electron, ε is the energy-loss and atomic units are used.

In the present measurements the momentum transfer $|\bar{k}| = |\bar{k}_0 - \bar{k}_1|$ spans over a wide range from 1. up to 80 a_0^{-1} .

The experiments on the Nitrogen compounds have been done in a crossed electron-gas beams apparatus, where the diffused electrons are selected in angle and analyzed in energy by a 180° hemispherical electron spectrometer. The analyzer can be rotated in the scattering plane from 0 to 130° with 0.1° precision and 0.5° resolution. The angular calibration has been checked by measuring the angular dependence both of the N. elastic cross-section, whose value is known from the literature (3), and of the KLL-Auger spectrum, which is expected to be isotropic. The achieved energy resolution FWHM was 0.7 eV when incident electrons of 1400 eV were used and 1.2 eV for $E_0 = 3400$ eV.

The relative differential cross section, measured on N_2 at several different angles, was brought to the absolute scale through the intensity ratio with the elastic cross section measured under the same experimental conditions. For the other molecules, for which the absolute elastic cross section is not known, the intensity of the $1s \rightarrow \pi^*$ resonance in N_2 has been used as calibration peak, by using gas mixture of known concentration as target. The experiment on BF_3 was performed in a different apparatus, having similar characteristics. Incident electrons of $E_0 = 1000$ eV kinetic energy were analyzed with $\Delta E = 1.1$ e FWHM, after the scattering with the BF_3 beam. The angle is known with .3° uncertainty.

THE $1s \rightarrow \pi^*$ TRANSITION IN N_2

The excitation of an inner K-shell electron in N_2 to the lowest unfilled valence state shows up with a very intense sharp transition few eV below the 1s orbital ionisation threshold. Depending on the spin and inversion symmetry of the final state, transitions from the ground $^1\Sigma$ state can in principle occur either to the singlet $^1\Pi_{u,g}$ or to the triplet $^3\Pi_{u,g}$ final states. The $^1\Sigma_g \rightarrow {}^3\Pi_{u,g}$ spin forbidden transition has been only recently observed by Shaw et al.(4) at a value $\varepsilon = 400.2$ eV in the energy-loss spectrum of electrons below 0.6 KeV.

The $^1\Sigma_g \rightarrow {}^1\Pi_{u,g}$ transition at $\varepsilon = 401.1$ eV was already well known. It has been largely investigated by photoabsorption (5-7) and high resolution energy-loss spectroscopy (8-12). The transition energy and the vibrational structure have been well established and several determinations of the OOS exist. Nevertheless, it is still debated whether the transition ends to a single broken symmetry state $(1s^{-1}2p)$ $^1\Pi$, or to two almost degenerate states: $(1\sigma_u^{-1})(1\pi_g)$ $^1\Pi_u$ and $(1\sigma_g^{-1})(1\pi_g)$ $^1\Pi_g$. Transitions to these states are respectively optically allowed and parity forbidden. Molecular symmetry considerations are in favor of the latter thesis, while to account for the experimental value of the transition energy, calculations must include relaxation effects and localisation of the core hole, leading to an equivalent core model and to a single allowed transition (3). The equivalent core model is also supported by high resolution studies of the vibrational structure of the electron energy loss peak, both at high and low momentum transfer (see for istance G.C. King in this book). On the other hand, the determinations of the OOS, as obtained from photoabsorption studies (6,7) or inferred by extrapolating the GOS measurement at small K's (9), span from f =0.12±.05 to f =0.23±.05 and do not exclude the contribution of optically forbidden transition to the GOS meausured by electron impact experiments.

In order to throw more ligth on this subject, the absolute value of $f(K,\varepsilon)$ for the singlet to singlet transition has been measured. The experimental results are reported in fig. 1. They have been obtained by applying the relation (1) to the cross section subtended by the resonance peak, after its deconvolution through the energy response of the spectrometer and the vibrational structure as measured by King et al.(10). For more details on the experimental conditions and the data analysis we refer to a forthcoming paper (14); here we will limit ourselves to comment the obtained results and to compare them with previous experiments and with the existing theoretical predictions.

The two sets of experimental data, taken at 1400 and 3400 eV, yield two independent determinations of the GOS, which are indistinguishable at least up to K=6 a_o^{-1} momentum transfer. This fulfils the necessity condition for the validity of the First Born Approximation, i.e. the independence of the GOS on variation of the incident energy. There is not direct evidence from the shape of the distribution of the presence of a forbidden transition showing up at large momenta. This transition is however suggested by the comparison with the calculated GOS. The only calculation, known to us, of the function $f(K,\varepsilon)$ for the resonance under study, is by Rescigno and Orel, (15) and on this book (13). This calculation, performed in a Configuration Interaction scheme, predicts two transitions to the $^1\pi_u$ and the $^1\pi_g$ states almost degenerate in energy (0.06 eV apart). The calculated $f(K,\varepsilon)$ for each separate transition is shown in fig. 1 by the full lines labelled respectively π_u and π_g. The energy resolution of the present experiment does not allow for resolving the two separate transitions; therefore comparison is done with the summed GOS $(\pi_g+\pi_u)$. The agreement between theoretical predictions and experimental findings is very good, but for a scaling factor (1.20) wich is indistinguishable from unit within two standard deviation of the quoted absolute error of the experiment. It is worth noting that contribution to the GOS from the forbidden (π_g) transition is essential in order to get agreement between theory and experiment for momentum transfer larger then 1 a_0^{-1}. The optically allowed transition alone gives an evaluation of the OOS close to the experimental value, but it predicts a wrong behaviour of the Oscillator Strength as a function of the momentum transfer.

For what concerns the OOS it has to be noted that it can not be directly measured by electron-impact spectroscopy, due to a minimum non zero momentum transferred in the scattering event. As a consequence the OOS derived in these experiments is always an estimated value from the measure of the GOS at small momentum transfer. To this purpose it is often used the following serial expansion (2):

$$f(K,\varepsilon)=f_0(\varepsilon) + K^2f_1(\varepsilon) + K^4f_2(\varepsilon) + \ldots\ldots$$

where the first term $f_0(\varepsilon)$ coincides with the OOS. This series, which for transitions into the continuum converges up to $K^2\leq\varepsilon$, has been applied as trial function in determining a least chi-squared best fit to our data. By this method the best f_0 obtained has been $f_0=.20\pm.02$. In table 1 this value is reported together with previous determinations both theoretical and experimental. It is consistent with the OOS recently calculated by Rescigno and Orel (13), and with the value by the

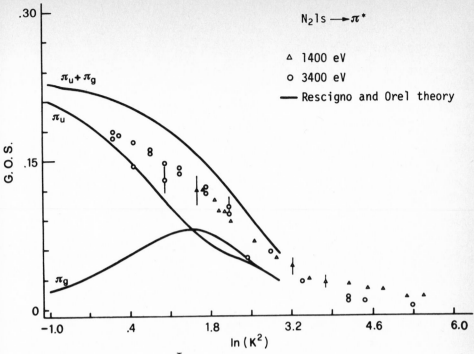

fig. 1 GOS for $1s \longrightarrow \pi^*$ excitation in N_2.

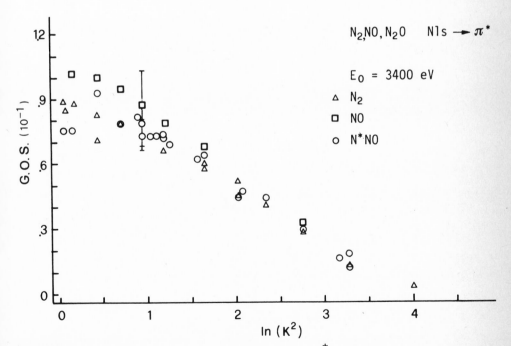

fig. 2 Comparison of total GOS for $1s \longrightarrow \pi^*$ excitation in N_2, NO and N^*NO.

earlier work of Dehmer and Dill (16). For what concerns the existin
experimental determinations, there is reasonable consistency amon
electron-impact results, while the optical determinations are consisten
tly lower, except for the x-ray absorption value by Bianconi et al. (7
which agrees with the present value within experimental uncertainties.

THE N1s $\longrightarrow$ nπ* TRANSITION IN NO AND N$_2$O

Also in NO and N$_2$O the energy-loss spectrum shows the presence of re
sonances close to the N K-edge, due to excitation of the N1s electron t
the unfilled π2p valence orbital (11,17). In NO the core excitatio
process brings the molecule from its ground $^2\Pi$ state to three fina
states with identical electronic configuration (N1s)$^{-1}$(π2p)^2 and diffe
rent symmetries: $^2\Sigma$, $^2\Delta$, $^2\Sigma^+$. The three transitions are unresolved i
the present experiment and result in a single broad peak (1.6 eV FWHM
in the energy-loss spectrum, having its centroid at 399.5 eV.

In N$_2$O, because of the different position occupied by the Nitroge
atoms in the molecule, the transition splits in two peaks 3.7 eV apar
from each other.

For the above mentioned transitions the relative GOS has been derive
following the same experimental procedure outlined for the N$_2$ measure
ments. Results obtained at 3400 eV incident energy are shown in fig.
together with the GOS for the correspondent transition in N$_2$. For N$_2$
the data actually refer to the peak at lower transition energy (hol
created in the external Nitrogen N*NO) because the intensity ratio wit
the NN*O transition has been found to be constant upon variation of th
momentum transfer. The three GOS have been suitably normalized in orde
to compare their shapes. Even though in eteronuclear molecules there i
no reason to introduce symmetry forbidden contributions, there is a mar
ked similarity among the shapes. This is also predicted by the recen
calculation by Rescigno and Orel (13). This result points to the rele
vance of the atomic character of the hole created, and reduces the in
fluence of the upper excited molecular orbital in the dynamics of th
process. The GOS is determined, indeed, by the projection of the core
hole wavefunction through the Coulomb operator e$^{i\,\overline{K}\cdot\overline{r}}$ over much more dif

fuse wavefunctions of the empty outer orbitals. The inner-hole state is essentially the same for the three molecules and, due to its localization it ends up in sampling the molecular excited charge over the N-site in a small region around the nucleus. In this picture it is the absolute value of the GOS, more than its K-dependence, to be affected by the chemical neighborhood of the inner-hole created, becoming a measure of the different electronegativity of the N atom when in different molecules. From the intensity ratios of the peaks respect to the N_2 resonance, we have derived the following ratios for the GOS per electron: $1.20\pm.18$ in NO, $1.04\pm.10$ in N^*NO, $1.32\pm.12$ in NN^*O. The GOS is indeed slightly higher when the Nitrogen is in presence of Oxygen.

We conclude that the GOS, in the N compounds investigated, is scarcely sensitive to the chemical neighborhood of the inner-shell created. Conversely the transition energy is known to change markedly upon change of the molecular structure where the N atom is located.

INNER-SHELL EXCITATION IN BF_3

Analogous investigation on polar molecules should better elucidate the relevance of the chemical enviroment to the inner shell excitations GOS. An experimental investigation on the BK* state in BF_3 has been recently started.
The Boron K-absorption spectrum of BF_3 is characterized by two intense non-atomic transitions respectively above and below the ionization threshold. Though these structures have been the subject of theoretical and experimental investigations, the origin for the broad strong band above the K-threshold is still an open question (18,19).

All the experiments up to date have been done in photoabsorption spectroscopy. In the present work the first results by high-energy electron energy-loss spectroscopy are reported. The energy-loss spectrum has been measured at several different scattering angles (from 2.8° up to 19.6°) with incident energy of 1000 eV. Two spectra are reported in fig.3 as measured at 2.8°(a) and 19.6°(b), after subtraction of a smooth continuum background. All the features observed in photoabsorption are present in the energy-loss spectra. The transition

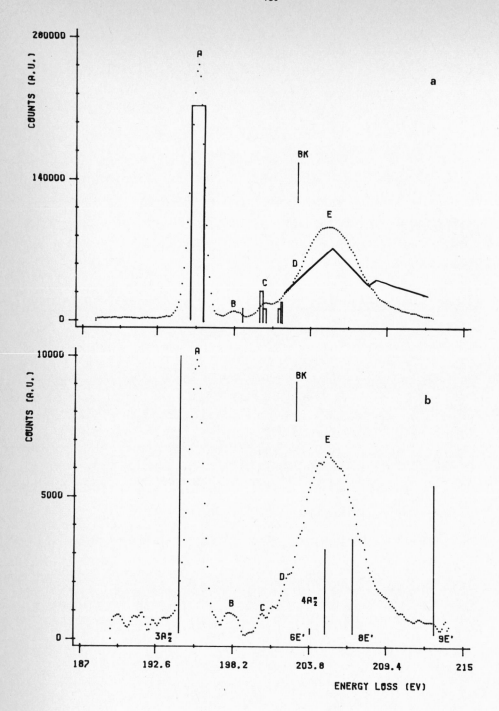

fig. 3 Electron energy-loss spectrum of B K-edge in BF$_3$ at 2.5°(a), 15.0°(b) scattering angle and 1000 eV incident energy.

energies of the two prominent peaks, which are interpreted as shape resonances, have been derived: the first sharp one lies at 195.5 eV, the second one wich is much broader has its centroid at 205.1 eV. The other weaker structures (peaks B,C,D in the figure) are observed at 198.1 eV, 200.5 and 200.9 eV respectively. They have been already interpreted as atomic-like transitions to Rydberg states. The energy positions are given with ±0.3 eV uncertainty. They are in very close agreement with the values obtained in the recent photoabsorption experiment by Ishiguro et al. (18). For a temptative assignement of the transitions the results of the calculations by these Authors and by Gianturco et al. (19) are reported in table 2 and in fig. 3. Even though the calculations refer to photoabsorption experiments there is good agreement also for the overall intensity distribution in the various regions of the spectra. It has, anyhow, to be noted that the relative intensity of the resonance peak E respect to the peak A becomes larger in the spectrum (b) at higher momentum transfer. This trend is confirmed by our measurements in the full range of angles investigated.

Further work is in progress along this line,with the aim of characterizing possible different contributions to this band on the basis of its momentum transfer distribution.

TABLE 1. Absolute Oscillator Strength for the transition $^1\Sigma_g \rightarrow (1s^{-1})(2p\pi)\,^1\Pi$

Ref.	experiment				theory			
	(6)*	(9)+	(7)*	this work	(16)	(13)	(20)	(21)
K	0	0.61	0	0(a)	0	0	0	0
f(K)	.12 +.05	.195 +.02	.23 +.06	.20 +.02	.23	.236	.257	.13

* photoabsorption experiment; + inelastic electron scattering
(a) extrapolated value, see text.

TABLE 2. Boron K-shell excitation in BF_3

Peak	experiment		theory					
	T.E. (eV)	T.E. (eV)	T.E. (eV)	OOS	state	T.E. (eV)	OOS	state
Ref.	(*)	(18)	(18)			(19)		
A	195.5	195.5	195.5	.1050	$2p\pi\,(a_2'')$	194.4	.0931	$3a_2''$
B	198.1	198.2	198.8		$3s+\sigma(a_1')$			
C	200.5	200.3	200.3	.0033	$3p\pi\,(a_2'')$	200.3	.0009	$2a_2''$
C			200.5	.0007	$3p\sigma\,(e')$	200.6	.0001	$5e'$
D	201.9	201.7	201.6	.0005	$4p\pi\,(a_2'')$			
D			201.7	.0008	$4p\sigma\,(e')$			
						203.9	.0018	$6e'$
						205.1	.0282	$4a_2''$
E	205.1	205.1	207.2	.116	$2p^*\sigma(e')$	205.2	.0001	$7e'$
						207.1	.0315	$8e'$
						212.9	.0496	$9e'$

(*) this work. Estimated energy uncertainty 0.3 eV

REFERENCES

1. King G C 1984 this book
2. Inokuti M 1971 Rev. Mod. Phys. $\underline{43}$ 297
3. Jansen R H J, de Heer F J, Luyken H J, van Vingerden B and Blaw H J 1976 J. Phys. B: At. Mol. Phys. $\underline{9}$ 185
4. Shaw D A, King G C, Read F H and Cvejanovic' D 1982 J. Phys. B: At. Mol. Phys. $\underline{15}$ 1785
5. Nakamura N, Sasanuma M, Sato S, Watanabe M, Yamashita H, Iguchi Y, Ejiri A, Nakai S, Yamaguchi S, Sagawa T, Nakai Y and Oshio T 1969 Phys. Rev. $\underline{178}$ 80
6. Wuilleumier F and Krause M O 1972 Int. Conf. on Inner-Shell Ionization Phenomena (Georgia Institute of Technology) p.773
7. Bianconi A, Petersen H, Brown F C and Bachrach R Z 1978 Phys. Rev. $\underline{A17}$ 1907
8. Wight G R, Brion C E and Van der Wiel M J 1972-1973 J. Electron Spectrosc. $\underline{1}$ 457
9. Kay R B. Van der Leuw Ph E and Van der Wiel M J 1977 J. Phys. B: At. Mol. Phys. $\underline{10}$ 2513
10. King G C, Read F H and Tronc M 1977 Chem. Phys. Lett. $\underline{52}$ 50
11. Tronc M, King G C and Read F H 1980 J. Phys. B: At. Mol. Phys. $\underline{13}$ 999
12. Hitchcock A P and Brion C E 1980 J. Electron Spectrosc. $\underline{18}$ 1
13. Rescigno T N and Orel A E 1984 this book
14. Camilloni R, Fainelli E, Petrocelli G and Stefani G to be published
15. Rescigno T N and Orel A E 1979 J. Chem. Phys. $\underline{70}$ 3390
16. Dehmer J L and Dill D 1976 J. Chem. Phys. $\underline{65}$ 5327
17. Wight G R and Brion C E 1974 J. Electron Spectrosc. $\underline{4}$ 313
18. Hishiguro E, Iwata S, Suzuki J, Mikuni A and Sasaki T 1982 J. Phys. B:At. Mol. Phys. $\underline{15}$ 1841
19. Gianturco F A, Semprini E and Stefani F 1983 Il Nuovo Cimento $\underline{2D}$ 687
20. Rescigno T N and Langhoff P W 1977 Chem. Phys. Lett. $\underline{51}$ 65
21. Arneberg R, Ågren H, Müller J and Manne R 1982 Chem. Phys. Lett. $\underline{91}$ 362

CORRELATION EFFECTS IN NEON STUDIED BY

ELASTIC AND INELASTIC HIGH ENERGY ELECTRON SCATTERING

J. J. McClelland and M. Fink
Physics Department
The University of Texas at Austin 78712

(Abstract)

New experimental investigations of correlation effects in neon
are presented. Elastic differential scattering cross sections are
measured with 0.1% precision, using 35 keV incident electrons.
Energy analysis is accomplished with a Møllenstedt analyzer. The
results, expressed in terms of a delta sigma curve and potential
energy quantities, show agreement with straightforward CI theories
and disagreement with Bethe–Goldstone predictions. This is in
contrast to previous measurements of Duguet et al.

I. Introduction

In the past 15 years, there have been significant advances made in the
calculation of atomic and molecular wave functions. Inclusion of correlation
effects has recently become more and more feasible with the improvement of
high speed computation facilities. These new theoretical results have led to
increased interest in experimental verification of the calculated wave
functions, with particular emphasis on the attempts to include correlation.
The standard benchmarks, like the total energy of the system, the dipole
moment, and the equilibrium internuclear distances in a molecule do provide a
check of whether the correlation is properly taken into account, but they
allow only limited conclusions from comparisons. Being just a single number
(or at best a few numbers), these quantities cannot contain full information
on the wave function. Experimental techniques which probe the entire wave
function are clearly superior if the most rigorous test is to be applied to
theory.

One such type of experiment is high energy electron scattering. Within the
framework of the first Born approximation, total, elastic and inelastic
electron scattering can be related to the scattering target by the Fourier
transform of the one- and two-electron charge densities, which reflect
directly on the wave function. Recent advances in the technology of high
energy electron scattering have made it possible to measure cross sections
with an accuracy of 0.1%, opening new possibilities for accurate
investigations emphasizing the correlation effects in the wave function. By
subtracting theoretical scattering intensities based on a Hartree–Fock (HF)
wave function from measured cross sections, one can create a difference curve
(referred to as a delta sigma curve) which shows exactly how the true wave
function differs from the HF. Plotting the difference calculated from the
same HF theory and theoretical correlated scattering intensities provides a
comparison which shows how far the new wave function has come towards a true
representation of the atom or molecule in question.

In the past, emphasis has been on the measurement of total scattering
intensities, since it does not require the added complication of energy
analysis in the detection system, and it already provides a fair amount of
information on correlation effects. We present here new results in which the
elastic contribution has been resolved, examining separately the first order

charge density. The scattering target chosen for this study was neon, since as a neutral closed shell atom it provides a good test case, for which a reasonable amount of theoretical work has been done. We compare the difference curve with two standard CI type theories[1,2] and a Bethe–Goldstone approximation.[3] Our results, in contrast to those of Duguet et al.[4] show the standard CI type wave functions as agreeing much better with experiment in the elastic channel than the Bethe–Goldstone approximation. Integration of the delta sigma curve and the ΔF curve is also carried out to calculate various components of the potential energy of the atom. Since these numbers are the result of subtracting two large numbers, large error bars have to be tolerated. Still, the results indicate that the CI results are preferred, in agreement with the recent results of Goruganthu and Bonham.[5]

II. Theoretical Background

This section contains a summary of the more relevant relations which allow correlation information to be extracted from elastic electron scattering measurements. The first Born approximation is assumed to be valid throughout the discussion, although in the actual data analysis it is circumvented in a way that will be discussed.

We first write, using atomic units, the elastic cross sections for an atom of atomic number Z as they relate to the x-ray coherent scattering factors F(K):[6]

$$\frac{d\sigma^{el}}{d\Omega} = 4K^{-4}(Z-F(K))^2 \tag{1}$$

where K is the momentum transfer ($K = 4\pi/\lambda \sin \theta/2$). As is well known in the theory of x-ray scattering, F(K) can be related to the charge density and hence the wave function of the atom through a Fourier transformation.

In order to show more clearly the effect of correlation on an atom, a delta sigma curve is created by subtracting partial wave scattering cross sections calculated from a relativistic HF wave function (corrected for exchange and spin flip contributions).

$$\Delta\sigma = \frac{K^4}{4}(I^{exp} - I_{PW}^{HF}) \tag{2}$$

It is in this process where the first Born approximation is corrected for. Since the data and the partial wave theory contain the same Born correction, cancellation will occur, leaving only the correlation effect as a visible difference between measurement and theory.

The first step in producing a delta sigma curve is to put the measured relative cross sections on an absolute scale. This is done in the elastic case by matching the data to the partial wave scattering intensities over a K range where no correlation effect is expected to be present. Then these theoretical cross sections are subtracted from the scaled data. Comparison can now justifiably be made with a non-relativistic theoretical curve calculated strictly in the Born approximation.

Besides looking at the shape of the delta sigma curve for comparison with theory, it is also possible to extract several potential energy quantities. These numbers can be compared with the results of Z-expansion theories[5], which bypass the wavefunction entirely, providing an independent check.

The elastic cross section can be related to the electron-nuclear part V_{ne} and the electron-electron part V_{ee} (coulomb) of the potential energy:

values larger than k = 1 a.u. nearly all the inelastic scattering is in the Compton profile, which centers around an energy loss of $E_o \sin\theta$, or 30 eV, whichever is larger.

The detector and monitor are identical to those in previous experiments done in this group.[8] They consist of RCA 8575 photomultipliers with pilot B scintillators, set up in a pulse counting mode. Phillips 474 300MHz discriminators allow high speed counting which is essential for low statistical error in a counting-type experiment. Relative angular position readout is provided by a Unitek Digisec 2^{19}-bit optical encoder with a resolution of 2 arc seconds. The angular position and detector and monitor signals, along with the digitized electron beam current, are stored in a Z80 microprocessor unit, which controls the experiment, and processed in the same manner or described in ref. 8. At each angular position, electrons (countrate ranging from 0.6 to 300 kHz) are counted for 300 seconds both with the gas nozzle on and with a background nozzle on. The detector signals are normalized by the monitor signals, the background signal (typically a factor of 50 to 100 smaller than the main signal) is subtracted, and zero angle is determined by an averaging routine, in which the two sides of data are folded on top of each other and the average center of symmetry is found. The resulting relative cross sections have an angular precision of about 2 arc seconds, and an intensity precision of the order of 0.1%.

IV.   Results and Discussion

Figures 2 and 3 show the elastic delta sigma curve and ΔF curve obtained from the data as described in section II. The HF wave function used to calculate the partial wave amplitudes was the relativistic one of Mann.[9] Also shown on the delta sigma plot are three theoretical calculations based on x-ray scattering factors obtained from Peixoto, Bunge and Bonham[1], Tanaka and Sasaki,[2] and Naon and Cornille[3]. The reference HF scattering factors used were the non-relativistic ones tabulated by Hubbell et al.[10]

The first two theories are based, respectively, on a 65 term CI expansion in which 85% of the correlation energy is reproduced, and a limited L-shell correlated wave function, where 55% of the energy is obtained. Both agree well with the measured results. The third theory, which is based on a Bethe-Goldstone (IEPA) approximation wave function yielding 97.3% of the correlation energy, shows a marked discrepancy when compared with the other theories and the experiment.

The elastic results shown here are in rather strong disagreement with the measurements of Duguet et al.[4], where the better agreement is found with the Bethe-Goldstone theory. We believe this discrepancy is attributable to their use of the Born approximation in producing the delta sigma curve. This hypothesis was tested by analyzing our data with pure Born amplitudes. The resulting delta sigma looked very similar to that of Duguet et al., having a slight positive bump at small K and a large negative dip around K = 4 a.u. It is unfortunate that this behavior is precisely what is predicted by the theory of Naon and Cornille.

Table I contains the various energy quantities derived from experiment. Also shown are the results of Duguet et al., along with the values obtained from the theoretical scattering factors. The experimental errors quoted, which appear to be large, are a result of noise in the data as well as a certain ambiguity in the matching process. Two trends can be seen in ΔV_{ne}: Large negative values for both are supported by the results of Duguet et al. and the Bethe-Goldstone calculations, while a small positive ΔV_{ne} is indicated by the Z-expansion theory and both CI calculations. Our results clearly agree with the latter, although the numerical value for ΔV_{ne} is somewhat larger than the theory predicts.

V. Conclusions

With the presentation of new data consisting of purely elastic electron scattering intensities, possibilities for detailed examination of calculated wavefunctions are opened. In the case of neon, it is seen that the measured elastic delta sigma curve shows marked agreement with one type of theory and strong disagreement with another. If these calculations were evaluated on the basis of correlation energy alone, the latter would have to be preferred, showing that electron scattering can contribute crucial information. We find that, despite the lack of full correlation, the standard CI-type theories of Peixoto et al. and Tanaka and Sasaki do better in predicting the form of the elastic delta sigma curve than does the Bethe-Goldstone-type wavefunction of Naon and Cornille. This conclusion is born out in the calculations of the various potential energy quantities, where additional support is provided by the Z-expansion theory of Bonham et al.

The extension of these techniques to molecules is the obvious next step, and is in progress now. Total measurements have already proven very useful for such molecules as N_2, CO_2, and CH_4[11], and it is hoped that resolution of elastic and inelastic scattering will provide additional insight.

Table 1   Correlation Effects of Potential Energy Quantities
(all quantities in eV)

	ΔV_{ne}	$\Delta V_{ne} + \Delta V_{ee}$(coulomb)
This work	+5.7±6	-2.5±6
Duguet et al.	-7.3±7	0.2±7
65-Term CI	+3.8	-3.3
L-shell CI	+2.3	-2.0
IEPA	-6.2	-1.4
Z-expansion	+1.4	-

VI. References

1. E. M. A. Peixoto, C. F. Bunge, and R. A. Bonham. Phys. Rev. 181 322 (1969).
2. K. Tanaka and F. Sasaki. Intl. Jl. Quantum Chem. 5 157 (1971).
3. M. Naon and M. Cornille. J. Phys. B 5 1965 (1972).
4. A. Duguet, A. Lahmann-Bennani and M. Roualt. J. Chem. Phys. 79 2786 (1983).
5. R. R. Goruganthu and R. A. Bonham. Phys. Rev. A 26 1 (1982).
6. R. A. Bonham and M. Fink. High Energy Electron Scattering (Van Nostrand Reinhold, New York, 1974).
7. H. F. Wellenstein, J. Appl. Phys. 44, 3668 (1973).
8. M. Fink, P. G. Moore and D. Gregory. J. Chem. Phys. 71 5227 (1979).
9. J. B. Mann (private communication).
10. J. H. Hubbell, W. J. Veigele, E. A. Briggs, R. T. Brown, D. T. Cromer and R. J. Howerton. J. Phys. Chem. Ref. Data 4 471 (1975).
11. M. Fink and P. G. Moore. Phys. Rev. A 15 112 (1977).

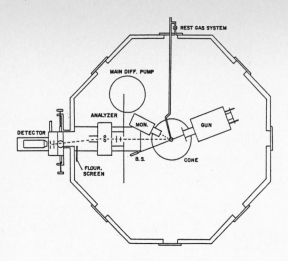

Fig. 1   High Energy Elastic Electron Scattering Apparatus
(top view).   Unit consists of an electron gun, a cone for
differential pumping, a monitor detector (Mon.), a beam stop (B.S.),
a Mollenstedt electron energy analyzer, a flourescent screen, a main
detector, and a rest gas system.

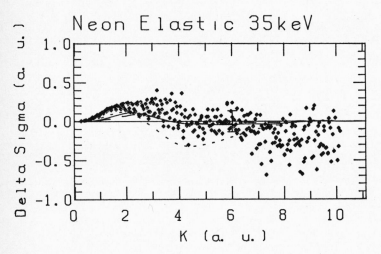

Fig. 2   Neon Elastic Delta Sigma (Partial Wave Thy.)
Experiment (◊), 65-term CI (——), L-shell CI (— —), Bethe-Goldstone (- - -).

$$V_{ne} + V_{ee} \text{ (coulomb)} = \frac{1}{\pi} \int (\frac{K^4}{4} I_{elast} - Z^2) dK$$

$$V_{ne} = \frac{2Z}{\pi} \int (\frac{K^2}{2} \sqrt{I_{elast}} - Z) dK$$

These relations cannot be exploited directly because the Born approximation has to hold on the absote scale to better than 0.1%. This problem can be circumvented by analyzing the difference between the experimental data and the partial wave theory. Similarly to $\Delta\sigma$, the change in $F(K)$ is derived as

$$\Delta F(K) = K^2 \cdot (\sqrt{I^{exp}} - \sqrt{I_{PW}^{HF}})/2$$

From $\Delta F(K)$ and $\Delta\sigma$ the quantities ΔV_{ne} and ΔV_{ee} (coulomb) can be determined by numerical integration.

III. Experimental

The data presented here represent the first results from the new electron diffraction unit at the University of Texas at Austin. Details of the apparatus will appear in a forthcoming publication; we mention here only the more important features.

A sketch of the set-up is shown in Fig. 1. The vacuum chamber is octagonal in shape, approximately 2 m in diameter and 1 m high, and has a base pressure of 2×10^{-6} torr. The walls are lined with μ-metal magnetic shielding and the laboratory is fitted with large Helmholtz coils to cancel the earth's magnetic field. The electron gun is of the telefocus type, operated at 35 kV and suspended from a rotating turntable. At the center of rotation is a 22 gauge stainless steel hypodermic needle, from which the target gas flows. Directly below this gas nozzle is an auxilliary diffusion pump fitted with a cone-shaped baffle reaching to within 1 cm of the nozzled tip. This arrangement greatly reduces the background scattering during the course of an experiment by maintaining a low residual pressure in the chamber. Research grade neon, obtained from the Matheson Co., was supplied to the nozzle through a regulated gas inlet system at a pressure of 14.6 torr. Also mounted on the turntable are a moveable beam stop and a monitor aimed at the scattering volume.

Attached to the wall of the chamber is a Möllenstedt electron energy analyzer constructed after the design of Wellenstein[7]. Two cylindrical rods 3/8" in diameter and 0.285" apart are placed between two grounded planes separated by 5". The rods are held at nearly the same voltage as the electron gun. A vertical slit allows scattered electrons to pass between the two rods slightly off center on the so-called caustic ray, where strong chromatic aberration causes energy dispersion. A detector is placed behind a second slit situated to catch only the elastic electrons. The resolution of the analyzer can be as small as 0.5 eV, with properly chosen slit widths, but in these experiments it was increased to approximately 20 eV by opening the exit slit. This large exit slit ensured a large angular acceptance for the analyzer, while not compromising the rejection of inelastic scattering. For momentum transfer

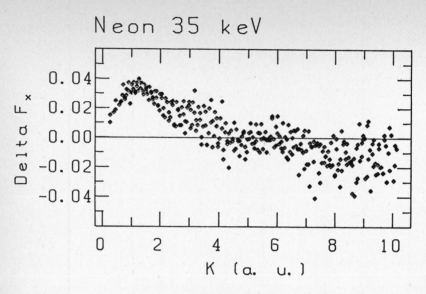

Fig. 3   Neon Delta F Curve.

COINCIDENCE MEASUREMENT OF THE ANISOTROPY OF $L_3M_{23}M_{23}(^1S_0)$ AUGER ELECTRONS OF ARGON

A. Lahmam-Bennani, G. Stefani*, A. Duguet

LCAM,[+] Université Paris-Sud, Bât. 351, 91405 ORSAY Cedex, FRANCE

*Laboratorio di Metodologie Avanzate Inorganiche del CNR, Via Salaria, 00016 Monterotondo Scalo, ROME, (ITALY)

Electron impact has long been used[1] to study the angular distribution of Auger electrons resulting from the electron rearrangement consecutive to an intermediate or inner shell ionization. In the case of an inner shell vacancy ($n\ell j$) with quantum number $j>1/2$, caused by impact ionization of a directed and unpolarized electron beam, the ion in general is left in a non-isotropic state. This anisotropy, or alignment of the ion, shows up in the anisotropy of the angular distribution of the decay products, namely the emitted x-ray photon or Auger electron.

Previous angular distributions have so far been measured[1] with only the Auger electron being detected, i.e. without detecting the scattered particle. Obviously, on thus averages over all directions and energies of the scattered electron, so that the anisotropy is very small, few percents at maximum, and is often difficult to measure. A recent theoretical study by Berezhko et al.[2] has shown that this anisotropy is much more pronounced when the Auger electron angular distribution is measured in coincidence with the fast scattered electron responsible of the inner hole creation. This unambiguously fixes the momentum transfer, so that the process is equivalent to the absorption of a polarized photon. The anisotropy of the angular distribution is due to the fact that the ionized atom "remembers" the direction of polarization of the incident photon.

Such an anisotropy is very sensitive first to the choice of the wavefunctions (which can hence be tested), but also and mostly to the description of the ionization process itself. The intermediate shells (such as the L-shell of argon) are of a prime interest as they are very sensitive to the inter- and intra-shell electron correlations, and will then give information about these correlations.

A coincidence experiment, performed on the argon $L_3M_{23}M_{23}(^1S_0)$ Auger transition has recently been reported by Sewell and Crowe[3], using 1 keV incident electrons. The fast scattered electron was detected at scattering angles, θ_S, of 15 and 21° (respectively corresponding to momentum transfer values K = 2.4 and 3.1 a.u.), and a mean ejected electron energy of 5 eV. At both angles, large disagreement was found with the first Born predictions of Berezhko et al [2].

On the other hand, Southworth et al[4] also reported a photo-absorption experiment on the $Xe-N_5O_1O_{23}$ ($^1P_1$) Auger transition, using synchrotron radiation. Qualitative agreement was there found with the Berezhko's et al predictions.

We report here on a similar electron impact study conducted on the Orsay electron-electron coincidence apparatus[5] where the Auger electron detection channel has been equipped by one of the hemispherical analyzers from the Rome electron-electron coincidence unit[6]. This allowed to increase the acceptance solid angle, $\Delta\Omega_A$, to $\sim 2.10^{-3}$ srd while keeping a sufficiently good energy resolution, $\Delta E_A \sim 1$ eV. Incident energy was 8255 eV, energy-loss of the scattered electron was ~ 5 eV larger than the Ar-2p ionization potential, while the scattering angles were $\theta_S = 1.5$ and 5.5 degrees (K = 0.5 and 2.4 a.u.), this last value being chosen such as to reproduce the K value used in Sewell and Crowe's work[3]. The scattered electron analyser was operated with an energy resolution $\Delta E_S \sim 10$ eV and a mean pass energy $E_S = 8000$ eV, while its acceptance angle was $\Delta\theta_S \sim\pm 0.2°$. For each fixed value of θ_S, an angular distribution was measured by recording the number of coincidences as a function of the Auger electron emission angle, θ_A, as defined with respect to the incident beam direction. θ_A could be varied from 20° to 115°.

In our experimental conditions, the maximum true coincidence rate was 0.2 count/sec (0.7 for the $L_3 M_{23} M_{23}(^3P_{012})$ line) with a true-to-random coincidence ratio of 1 to 3, leading to a 5% statistical error after $\sim$ 12 hrs accumulation time. This is to be compared to the maximum coincidence rate of 0.01 count/sec and the 15–50% statistical error reported by Sewell and Crowe.

Extensive tests have been carried out in order to rule out any possible experimental problem. The kinematic conditions ensured that the contribution to the measured coincidence signal from (e,2e) –type reactions was negligible[7] The coincidence angular distribution of the Ar-$L_3 M_{23} M_{23}$ ($^3P_{012}$) transition was measured and, as expected, it was found to be isotropic within statistical uncertainties. The (e,2e) coincidence angular distribution of secondary electrons ejected from He with the same energy as the $L_3 M_{23} M_{23}$ (^1S_o) Auger ones (i.e. 201 eV) was measured in impulsive (or binary) conditions (θ_S = 9.5°). The corresponding absolute experimental triple differential cross sections (TDCS) are shown in figure 1.

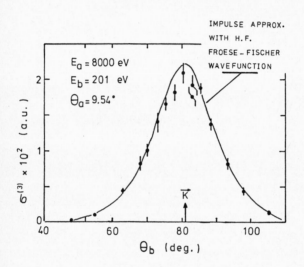

Fig. 1 : Experimental and theoretical (e,2e) TDCS for He. The two outgoing electrons have the same final energies as in the Auger angular distributions of Fig. 2.

They are in very good agreement with the theoretical ones calculated using the "corrected" impulse approximation[5] and a Hartree-Fock-Froese-Fischer He ground state wavefunction.

The angular distributions of the $L_3 M_{23} M_{23}$ (^1S_o) Auger electrons measured at θ_S = 1.5° and 5.5° are shown in figure 2 a and b, respectively. At K = 0.5 a.u., a less than 10% anisotropy is observed with a minimum in the momentum transfer direction, $\vec{K}$ (θ_A = 64° in our case) as predicted by Berezhko et al's 1st Born theory. Indeed the first Born approximation requires any angular distribution to be symmetric around $\vec{K}$. At the larger K value , K = 2.4 a.u., the observed anisotropy exceeds 50 % in the investigated angular range and is much larger than given by Berezhko et al's theory. Also the minimum in the angular distribution is shifted from the $\vec{K}$ direction (θ_A = 77°) towards larger angles, similar to the observations reported by Sewell and Crowe at the same K value and lower impact energy, E_o.

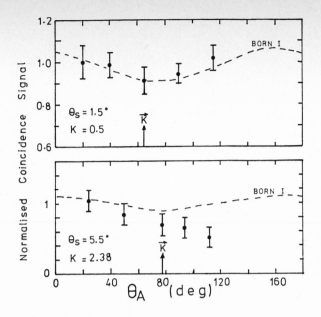

Fig. 2 : Coincidence angular distribution of the $L_3 M_{23} M_{23}$ ($^1S_0$) Auger electrons.

To try to ascribe which of the two parameters, K or E_0, is responsible of this failure of the first Born approximation, experiments will be carried out in the immediate future at an intermediate incident energy and the same momentum transfer values. Further studies along this line are also under way, specially on the M_4 $N_{23} N_{23}$ ($^1S_0$) Auger electrons of Krypton ($3d_{3/2}$ vacancy state).

[†](laboratoire associé au CNRS N° 281)

References

1. B. Cleff, W. Mehlhorn J.Phys.B 7, 605 (1974)
2. E.G. Berezhko, N.M. Kabachnik and V.V. Sizov J.Phys.B 11, 1819 (1978)
3. E.C. Sewell, A. Crowe J.Phys.B 15, L357 (1982)
4. S.H. Southworth, P.H. Kobrin, C.M. Truesdale, D. Lindle, S. Owaki and D.A. Shirley Phys. Rev. A24, 2257 (1981)
5. A. Lahmam-Bennani, H.F. Wellenstein, C. Dal Cappello, M. Rouault and A. Duguet J.Phys.B 16, 2219 (1983)
6. A. Giardini-Guidoni, R. Tiribelli, D. Vinciguerra, R. Camilloni, G. Stefani and G. Missoni "Momentum wavefunctions 1976", AIP Conference Proceedings, Ed. by D.W. Devins, p. 205
7. An additional argument is that the intensities of the $L_3 M_{23} M_{23}$($^3P_{012}$) and $L_3 M_{23} M_{23}$($^1S_0$) lines were found to be in the same ratio as the observed coincidence rates between the scattered electron and each of these lines. This indicated that the background intensity under the Auger lines did not contribute significantly to the coincidence signal.

PART V

MOLECULAR EXCITATION BY ELECTRON

IMPACT: THE THEORETICAL TREATMENTS

ELECTRONIC EXCITATION OF MOLECULES

BY ELECTRON IMPACT

Vincent McKoy and Mu-Tao Lee*

Arthur Amos Noyes Laboratory of Chemical Physics
California Institute of Technology
Pasadena, California 91125 USA

Introduction

In this talk I will review the recent progress that has been made in the theoretical determination of differential and integral cross sections for the electronic excitation of molecules by low-energy electrons. Whereas there has been considerable progress in the development and application of theoretical methods for treating inelastic electron-atom scattering [1], the situation is quite different for the related molecular problem. It is well-known that this difference is ultimately due to the difficulties associated with the nonspherical nature of the electron-molecule force field. In fact, it was not until the 1960's that a renewed interest in processes involving H_2 and N_2 resulted in the application of plane-wave theories to treat the excitation of electronic states in these molecules. In the last five years more advanced theories such as the distorted-wave method, the impact-parameter method, and the close-coupling method have been applied to the description of electron impact excitation of diatomic molecules. The results of these more recent applications, and their comparison with available experimental data, will be the focal point of my presentation. We will see that, although the results of some of these theoretical applications are encouraging, there are substantial disagreements between both the predictions of the different methods themselves, and with the experimental data. There clearly remains a serious need for further development of theoretical methods for the prediction of electron impact excitation cross sections of both linear and polyatomic molecules.

In this discussion of inelastic cross sections the emphasis will be on the differential cross sections since these cross sections provide a more meaningful test of the collision physics of the process than do integral cross sections. Such comparisons are important since they show that lowest-order theories, e.g., the Ochkur-Rudge approximation, generally predict qualitatively incorrect differential cross sections and that any quantitative agreement between the corresponding integral cross sections and experiment is likely to be fortuitous. Moreover, these comparisons show that the distorted-wave approximation seems to contain the minimum physics required to predict differential cross sections in qualitative agreement with experiment [2]. Many of these comparisons will be drawn from the extensive set of differential and integral cross sections which have been determined for N_2. In fact, a survey of the cross section data for electron impact excitation of molecules shows that this is the only molecule for which these differential and integral cross sections have been extensively studied [2]. In view of the important role that electron impact processes play in several phenomena, it is obvious that further experimental studies of these cross sections for other molecules are needed.

Finally, we will concentrate completely on direct excitation processes and we will not discuss resonant excitation cross sections. Although core-excited resonances can cause significant increases in these cross sections, they do so over an energy region small compared to the region over which direct excitation is important.

Summary of Methods

I will begin by making some pertinent comments about the various theoretical approaches which have been used to study electron impact excitation cross sections of molecules. These approaches fall into four categories:

- plane-wave theories such as the Born and Ochkur-Rudge approximations
- the impact-parameter method
- distorted-wave theories
- multichannel theories.

Plane-wave theories are lowest-order theories which contain only the collision physics appropriate to high electron energies and hence predict qualitatively and quantitatively incorrect differential cross sections at low and medium energies. However, these theories continue to be used in the low and medium energy region because they are easy to apply computationally and, even though the differential cross sections are generally qualitatively incorrect, the integral cross sections can appear reasonable even at medium energies [2]. These theories have been widely applied to obtain electron impact excitation cross sections of molecules. Some examples of molecular systems along with the electronic bands to which these theories have been applied include

- H_2 $X\ ^1\Sigma_g^+ \to b\ ^3\Sigma_u^+,\ a\ ^3\Sigma_g^+,\ B\ ^1\Sigma_u^+,\ C\ ^1\Pi_u,\ c\ ^3\Pi_u,\ e\ ^3\Sigma_u^+,$ and others.

  See Refs. [3], [4], [5] and elsewhere.

- N_2 $X\ ^1\Sigma_g^+ \to A\ ^3\Sigma_u^+,\ B\ ^3\Pi_g,\ W\ ^3\Delta_u,\ B'\ ^3\Sigma_u^-,\ a\ ^1\Pi_g,\ W\ ^1\Delta_u,\ b'^1\Sigma_u^+,\ c'^1\Sigma_u^+,$

  $C\ ^3\Pi_u$, and others. See Refs. [6] and [7].

- CO $X\ ^1\Sigma^+ \to a\ ^3\Pi,\ A\ ^1\Pi,\ d\ ^3\Delta,\ D\ ^1\Delta,\ a'^3\Sigma^+,\ B\ ^1\Sigma^+,\ C\ ^1\Sigma^+,$ and others.

  See Ref. [8].

- O_2 $X\ ^3\Sigma_g^- \to B\ ^3\Sigma_u^-$. See Ref. [9].

These results, along with the results of the other methods mentioned above and the available experimental data, will be discussed collectively in the next section.

The impact-parameter method is a semiclassical approach to electron-molecule collisions in which the molecular electrons are treated quantum mechanically while the scattered electron is treated purely classically using straight-line trajectories [10]. The method was developed originally to describe electronic excitation of atoms [11] and is especially suited for electric dipole transitions. The calculation of the cross sections by this method requires an integration over impact parameters and, in fact, a cut-off must be introduced in this integration so as to exclude the divergent contributions arising from small impact parameters [10]. The impact parameter method must certainly be viewed as a low-order treatment of the continuum physics in electron-molecule collisions. Molecular applications have shown that the method can be an improvement over the Born approximation at lower impact energies [10]. In these molecular applications no differential inelastic cross sections were reported. For reasons mentioned above, it is unlikely that the differential cross sections will be qualitatively or quantitatively correct at these lower impact energies. The impact parameter method has been used to study the excitation of the following molecular band systems [10]:

- H_2 $X\ ^1\Sigma_g^+ \to B\ ^1\Sigma_u^+$ from 20 to 100 eV.
- N_2 $X\ ^1\Sigma_g^+ \to b'^1\Sigma_u^+$ and $c'^1\Sigma_u^+$ from 20 to 150 eV.
- F_2 $X\ ^1\Sigma_g^+ \to 1\ ^1\Sigma_u^+$ and $3\ ^1\Pi_u$ from 15 to 100 eV.

Studies of inelastic electron scattering from light atoms [1,2,13] have shown that distorted-wave methods have considerable utility in the intermediate energy region extending from the ionization threshold up to several hundred eVs and that they are a substantial improvement over plane-wave theories, particularly at lower energies.

Although more approximate than close-coupling methods, the distorted-wave approximation is easier to apply. especially to molecules, and is the next logical step beyond the plane-wave approximation. The development of methods for the determination of the necessary elastic continuum wave functions of electron-molecule systems has clearly made possible the application of distorted-wave methods to electronic excitation of linear molecules to date [14-21].

It is necessary to discuss a few relevant details of the distorted-wave approximation which we have used to obtain the electronically inelastic cross sections for H_2, N_2, F_2, CO, and CO_2. Within the Born Oppenheimer and Franck-Condon approximation, and assuming the target rotational levels to be degenerate, the differential cross section for electronic excitation by electron impact can be written as

$$\frac{d\sigma}{d\Omega} (n \leftarrow 0;E,\hat{r}') = SM_n \sum_{\nu'} \frac{k_{\nu'}}{k_0} q_{\nu'0} \frac{1}{8\pi^2} \int d\hat{R}' |f_{k_0} (n \leftarrow 0;R',\hat{r}')|^2 \tag{1}$$

for impact energy $E = \frac{1}{2} k_0^2$. In Eq.(1) $f_{k_0} (n \leftarrow 0,R',\hat{r}')$ is the fixed-nuclei scattering amplitude in the laboratory frame, R' denotes the nuclear coordinates of the target, and $\hat{r}'$ denotes the scattering angles. The symbol $q_{\nu'0}$ is the Franck-Condon factor between the $\nu=0$ ground state vibrational level and the ν' level of the excited state and k_0 and $k_{\nu'}$ are the momenta of the incoming and outgoing electron respectively. The factor S results from summing over final and averaging over initial spin sublevels and equals $\frac{1}{2}$ for singlet to singlet excitation and 3/2 for singlet to triplet excitation [15]. For a linear molecule M_n is the orbital angular momentum projection degeneracy factor of the final electronic state, e.g., 1 for a Σ state and 2 for a Π state.

The scattering amplitude is treated in a distorted-wave approximation which can be readily derived from the two-potential formula [22]. In the present applications the initial target state is the Hartree-Fock ground state and the final target state is treated in the single-channel Tamm-Dancoff approximation (TDA) [15]. This single-channel TDA is equivalent to an independent-electron picture in which the excited orbital is an eigenfunction of the V_{N-1} potential due to the N-1 core electrons [23]. Moreover, both the incident and outgoing electron distorted wave functions are calculated in the Hartree-Fock field of the ground state. It is important to note that this distorted-wave model differs from the usual distorted-wave approximation in which the scattered electron moves in the field of the final state [12,24,25]. We note that if one used these same incident and outgoing electron distorted wave functions but treats the target electronic transition density in the random phase approximation, the resulting distorted-wave model is entirely equivalent to the first-order many-body perturbation theory formulation of electron impact excitation of atoms and molecules [12,22].

In this formulation the electronic portion of the transition matrix in the body-fixed frame can be shown to involve matrix elements of the form [15]

$$< k_n,n|T_{el}|k_00 > = < \phi_n \psi_{k_n}^{(-)} |\upsilon|\phi_\alpha \psi_{k_0}^{(+)} >_a \tag{2}$$

where $\psi_{k_0}^{(+)}$, $\psi_{k_n}^{(-)}$ are initial, final Hartree-Fock (static-exchange) continuum spin-orbitals satisfying outgoing-wave, incoming-wave boundary conditions; ϕ_α is an occupied Hartree-Fock spin-orbital and ϕ_n is a spin-orbital of the V_{N-1} potential formed by removing an electron from the target orbital α. The antisymmetrized matrix element is defined as

$$< ij|\upsilon|k\ell >_a = < ij|\upsilon|k\ell > - < ij|\upsilon|\ell k > \tag{3}$$

where

$$< ij|\upsilon|k\ell > = \int dx_1\, dx_2\, \phi_2^*(x_1)\, \phi_j^*(x_2)\, \frac{1}{r_{12}}\, \phi_k(x_1)\, \phi_\ell(x_2). \tag{4}$$

A partial-wave expansion of the continuum orbitals, ψ_{k_0} and ψ_k, is used to obtain a single-center expansion of the transition matrix in the $\underset{\sim}{k}_n$ body-fixed frame from which the laboratory frame scattering amplitude can be obtained directly. This scattering amplitude is then inserted in Eq.(1) and the orientational averaging carried out. In most of these applications the elastic continuum orbitals, ψ_{k_0} and ψ_{k_n}, are obtained by solving the Lippmann-Schwinger equations using the iterative Schwinger variational procedure [26].

The results of applications to this distorted-wave theory to electronic excitation of molecules by electron impact show that the model does quite well in reproducing the important features in the differential cross sections. This form of the distorted-wave approximation thus appears to contain the minimum physics required to predict differential cross sections in qualitative agreement with experiment. This distorted-wave approximation has been used to study the excitation of the following band systems:

- H_2 $X\,^1\Sigma_g^+ \to b\,^3\Sigma_u^+$, $B\,^1\Sigma_u^+$, $C\,^1\Pi_u$, $c\,^3\Pi_u$, $B'^1\Sigma_u^+$, and $E(F)^1\Sigma_g^+$.

  See Refs. [14,15,18].

- N_2 $X\,^1\Sigma_g^+ \to A\,^3\Sigma_u^+$, $w\,^1\Delta_u$, $W\,^3\Delta_u$, $a\,^1\Pi_g$, $B\,^3\Pi_g$, $C\,^3\Pi_u$, $E\,^3\Sigma_g^+$, $b'^1\Sigma_u^+$,

  and $c'^1\Sigma_u^+$. See Refs. [16,19].

- CO $X\,^1\Sigma^+ \to A\,^1\Pi$, $a\,^3\Pi$, $a'^3\Sigma^+$, $d\,^3\Delta$, and $D\,^1\Delta$. See Ref. [20].

- F_2 $X\,^1\Sigma_g^+ \to 1\,^2\Pi_u$. See Ref. [17].

- CO_2 $X\,^1\Sigma_g^+ \to 1,^3\Sigma_u^+$, $1,^3\Pi_g$, $1,^3\Pi_u$, and $1,^3\Delta_u$. See Ref. [21].

Multichannel theories of electronic excitation by low-energy electrons are just beginning to be applied to molecular targets. The complexity of molecular force fields obviously make the problems associated with the application of multichannel theories to molecular targets substantially more severe than for atomic targets. However, two-state close-coupling calculations have been carried out for selected transitions in H_2 by Chung and Lin [27] and by Weatherford [28] and for excitation of the a $^1\Pi_g$ state in N_2 by Holley et al. [29]. The integral cross sections obtained by these close-coupling calculations agree quite well with those of the distorted-wave studies for excitation of the B $^1\Sigma_u^+$ state in H_2 [15,27]. The corresponding cross sections for excitation of the b $^3\Sigma_u^+$ state, however, are in poor agreement [15,27].

More recently, we have used a multichannel extension of the Schwinger variational principle to study the cross sections for excitation of the b $^3\Sigma_u^+$ of H_2 at the two-channel level. Details of this multichannel formulation have been discussed elsewhere [30,31]. An important feature of this formulation, however, is that it is designed to be applicable to molecules of arbitrary geometry and to allow for the inclusion of a substantial number of open and closed channels [32].

In summary, multichannel theories of electronic excitation by electron impact have been applied to the following molecular systems:

- Two-state close-coupling calculations:

  - H_2 $X\,^1\Sigma_g^+ \to b\,^3\Sigma_u^+$, $B\,^1\Sigma_u^+$, $a\,^3\Sigma_g^+$, $C\,^3\Pi_u$, and $e\,^3\Sigma_u^+$. See Ref.[27,28].

  - N_2 $X\,^1\Sigma_g^+ \to a\,^1\Pi_g$. See Ref. [29].

O Multichannel Schwinger method at the two-channel approximation:

o H_2 $X\ ^1\Sigma_g^+ \to b\ ^3\Sigma_u^+$. See Ref. [32].

Applications and Discussion

Figure 1 compares the calculated differential cross sections for the $X\ ^1\Sigma_g^+ \to b\ ^3\Sigma_u^+$

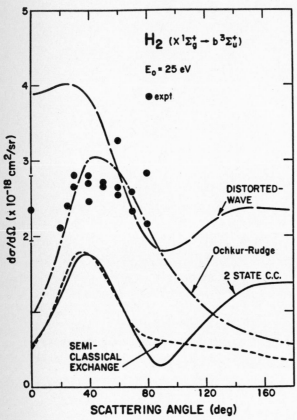

H_2 $(X^1\Sigma_g^+ - b^3\Sigma_u^+)$

$E_0 = 25$ eV

● expt

DISTORTED-WAVE

Ochkur-Rudge

2 STATE C.C.

SEMI-CLASSICAL EXCHANGE

$d\sigma/d\Omega$ $(\times 10^{-18}$ cm^2/sr)

SCATTERING ANGLE (deg)

Fig. 1 Differential cross section for excitation of the b $^3\Sigma_u^+$ state of H_2 at 25 eV. The experimental data are relative and have been normalized to the Ochkur-Rudge results in a least-squares sense [32]. The distorted-wave results are from Ref. 15 and the two-state close-coupling results, including those obtained with a semiclassical exchange potential, are those of Ref.28. This figure is taken from Ref. 2.

excitation in H_2 at 25 eV with the relative experimental data of Trajmar et al.[33]. We note that the experimental data is relative and has been least-squares normalized to the Ochkur-Rudge cross sections because no other absolute scale was available at that time. Although the close-coupling and distorted-wave cross sections are very different in magnitude, their shapes are quite similar at angles beyond 35°. The monotonically decreasing cross sections of the Ochkur-Rudge theory for angles beyond 90° is characteristic of a low-order model. This behavior will be seen in many other transitions.

In Fig. 2 we show the integral cross sections for excitation of the b $^3\Sigma_u^+$ state in H_2 obtained by the distorted-wave method, the Ochkur-Rudge theory, and the two-state close-coupling method [27] along with the experimental data of Corrigan [34] as analyzed by Cartwright and Kuppermann [3]. These measured cross sections should be an upper limit to the direct excitation cross section for the b $^3\Sigma_u^+$ state. The distorted-wave cross sections are substantially larger than those of the close-coupling calculations. When cascade contributions from other triplet states to the

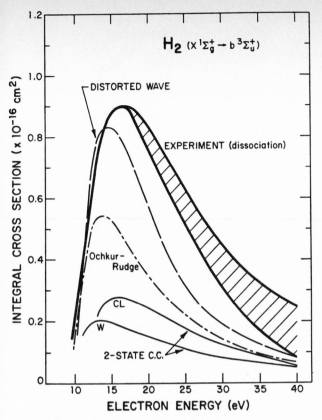

Fig. 2 Integral cross sections for excitation of the b $^3\Sigma_u^+$ in H_2. The experimental data were determined as cross sections for dissociation of H_2 into neutral fragments. The distorted-wave cross sections are from Ref. 15, the Ochkur-Rudge results from Ref. 3, and the two-state close-coupling results from Refs. 27 and 28, CL and W respectively. This figure is taken from Ref. 2.

dissociation process are included, the resulting total cross section for dissociation is in reasonable agreement with the experimental data [27], suggesting that the close-coupling cross sections for the b $^3\Sigma_u^+$ excitation are reasonable and that the distorted-wave approximation has seriously overestimated this cross section. We will see that results of multichannel Schwinger calculations confirm this conclusion [32].

The cross sections for the X $^1\Sigma_g^+ \to$ b $^3\Sigma_u^+$ excitation have also been studied at the two-channel level using a multichannel extension of the Schwinger variational principle [30-32]. In Table 1 we compare the results of these calculations with those of

Table 1 Integral cross sections in 10^{-17} cm^2 for the X $^1\Sigma_g^+ \to$ b $^3\Sigma_u^+$ excitation in H_2

Energy(eV)	DW[a]	CC[b]	SMC[c]
15	8.30	2.80	3.42
20	5.78	2.53	2.97
30	1.95	1.26	1.18
40	0.82	0.62	0.55

[a] distorted-wave results of Ref. 15; [b] two-state close-coupling studies of Ref. 27; [c] two-state Schwinger multichannel calculation of Ref. 32.

the two-state close-coupling studies and the distorted-wave approximation. The agreement between the close-coupling results [27] and the Schwinger multichannel calculations is encouraging over this energy range and suggests that the distorted-wave approximation has overestimated these cross sections.

Figure 3 shows the differential cross sections for the $X\ ^1\Sigma_g^+ \to B\ ^1\Sigma_u^+$ (v' = 2) excitation obtained by the Born and distorted-wave approximations at 15 and 40 eV

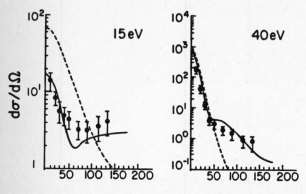

Fig. 3 Differential cross section (in 10^{-20} cm^2/sr) for excitation of the B $^1\Sigma_u^+$ (v'=2) state of H$_2$ at 15 eV and 40 eV. The experimental data is from Ref. 35 and the solid and dashed lines are from the distorted-wave and Born approximations respectively [15].

along with the experimental data. The Born differential cross sections agree poorly with the experimental results for all angles at 15 eV and for scattering angles beyong 50° at 40 eV and other energies [15]. The differential cross sections of the distorted-wave method are in good qualitative agreement with the experimental data. The model, however, overestimates the cross section in the forward direction. The close-coupling studies did not report any differential cross sections [27]. Figure 4 contains a comparison of the measured and various theoretical integral cross

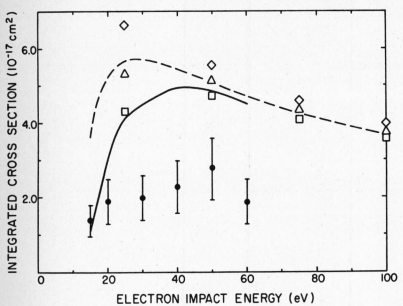

<u>Fig. 4</u> Cross section for excitation of the B $^1\Sigma_u^+$ state of H$_2$; ———: distorted-wave results [15];— — —: Born results [15]; □: close-coupling results [27]; Δ: Ochkur-Rudge results [27]; ◇ : Born results [27]; ●: experimental data [35].

sections for excitation of the B $^1\Sigma_u^+$ state of H_2. The close-coupling and distorted-wave cross sections agree quite well for this dipole-allowed excitation. Both sets of cross sections are significantly larger than the experimental cross sections over most of the energy range. The source of this disagreement is not understood and further experimental studies of this transition may be useful. For this transition the cross sections obtained by the impact parameter method [10] agree quite well with those of the distorted-wave studies.

Figures 5 and 6 compare the experimental differential cross sections with those of the Ochkur-Rudge and distorted-wave methods for excitation of the A $^3\Sigma_u^+$ and C $^3\Pi_u$ states of N_2 at 20 and 50 eV. The Ochkur-Rudge method does poorly in predicting these cross sections while the distorted wave model does quite well in reproducing the shapes of these differential cross sections. The magnitudes of these distorted-wave cross sections are, however, too large at most scattering angles. These, and other results, show that the distorted-wave method generally provides the qualitatively correct differential cross sections for exchange transitions [2,19]. Figures 7 and 8 show the differential cross sections for excitation of the c' $^1\Sigma_u^+$ and a $^1\Pi_g$ states of N_2 at 60 and 30 eV. The agreement between the distorted-wave and experimental results for excitation of the a $^1\Pi_g$ state are encouraging in view of the importance of both direct and exchange contributions in this excitation [19].

Extensive comparisons of the calculated and experimental integral cross sections for excitation of many electronic states of N_2 have been made recently [2,19]. As stated earlier, agreement between the calculated and measured integral cross sections does not always imply that the theoretical approach is necessarily reproducing the essential physics of the collisions. Figure 9 compares the calculated and experimental integral cross sections for excitation of the c' $^1\Sigma_u^+$ state of N_2. Electron correlation effects are important in the determination of these cross sections. [10,19].

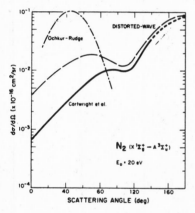

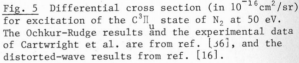

Fig. 5 Differential cross section (in 10^{-16} cm^2/sr) for excitation of the $C^3\Pi_u$ state of N_2 at 50 eV. The Ochkur-Rudge results and the experimental data of Cartwright et al. are from ref. [36], and the distorted-wave results from ref. [16].

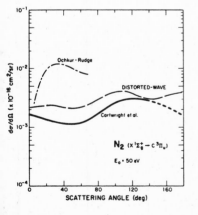

Fig. 6 Differential cross section (in 10^{-16} cm^2/sr) for excitation of the A $^3\Sigma_u^+$ state of N_2 at 20 eV. The Ochkur-Rudge results and the experimental data of Cartwright et al. are from ref. [36] and the distorted-wave results from ref. [19].

204

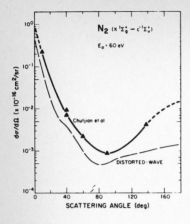

Fig. 7 Differential cross section (in 10^{-16} cm^2/sr) for excitation of the c' $^1\Sigma_u^+$ state of N$_2$ at 50 eV. The experimental data of Chutjian et al. is from ref. [37] and the distorted-wave results from ref. [19].

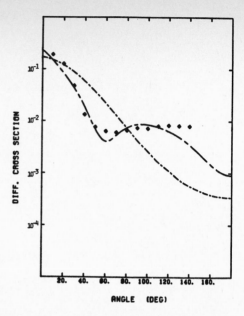

Fig. 8 Differential cross section (in 10^{-16} cm^2/sr) for excitation of the a$^1\Pi_g$ state of N$_2$ at 30 eV. ———— ———: distorted-wave results [19]; ———— Born results [19]; ♦: experimental results of ref. [36].

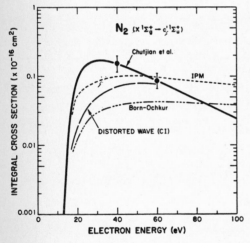

Fig. 9 Integral cross sections for excitation of the c' $^1\Sigma_u^+$ state of N$_2$. The experimental results of Chutjian et al. are from ref. [37], the Born-Ochkur results from ref. [6], the impact-parameter results from ref. [10], and the distorted-wave results from ref. [19].

Acknowlegements

This material is based upon work supported by the National Science Foundation under Grant No. PHY-8213992 and Grant No. INT-8219691 (Office of International Programs, U.S.-Brasil Program). The authors acknowledge computing support from the National Center for Atmospheric Research (NCAR) which is sponsored by the National Science Foundation.

*Permanent address: Departamento de Quimica, Universidade Federal de São Carlos, C.P. 676, São Carlos, S.P. Brasil.

References

1. B. H. Brandsen and M.R.C. McDowell, Phys. Rpts. 46, 249 (1978).
2. S. Trajmar and D.C. Cartwright, in Electron-Molecule Interactions and Their Applications, edited by L.G. Christophorou (Academic, New York, 1983), Chap. 2.
3. D.C. Cartwright and A. Kuppermann, Phys. Rev. 163, 86 (1967).
4. S. P. Khare, Phys. Rev. 149, 33 (1966).
5. S. Chung, C.C. Lin, and E.T.P. Lee, Phys. Rev. A 12, 1340 (1975).
6. S. Chung and C. C. Lin, Phys. Rev. A 6, 988 (1972).
7. D.C. Cartwright, Phys. Rev. A 2, 1331 (1970); ibid A 5, 1974 (1972).
8. S. Chung and C. C. Lin, Phys. Rev. A 9, 1954 (1974).
9. S. Chung and C. C. Lin, Phys. Rev. A 21, 1075 (1980).
10. A. U. Hazi, Phys. Rev. A 23, 2232 (1981).
11. M. J. Seaton, Proc. Phys. Soc. (London) 79, 1105 (1962).
12. L. D. Thomas, G. Csanak, H. S. Taylor, and B.S. Yarlagadda, J. Phys. B 7, 1719 (1974).
13. N.T. Padial, G.D. Meneses, F. J. da Paixão, G. Csanak, and D.C. Cartwright, Phys. Rev. A. 23, 2194 (1981).
14. T.N. Rescigno, C.W. McCurdy,Jr., V. McKoy, and C. F. Bender, Phys. Rev. A 13, 216 (1976).
15. A. W. Fliflet, and V. McKoy, Phys. Rev. A 21, 1863 (1980).
16. A. W. Fliflet, V. McKoy, and T.N. Rescigno, J. Phys. B. 12, 3281 (1979).
17. A. W. Fliflet, V. McKoy, and T. N. Rescigno, Phys. Rev. A 21, 788 (1980).
18. M. T. Lee, R. R. Lucchese, and V. McKoy, Phys. Rev. A 26, 3240 (1982).
19. M. T. Lee, and V. McKoy, Phys. Rev. A 28, 697 (1983) and ibid (to be published) 1984.
20. M. T. Lee and V. McKoy, J. Phys. B 15, 3971 (1982).
21. M. T. Lee and V. McKoy, J. Phys. B 16, 657 (1983).
22. T. N. Rescigno, C. W. McCurdy, Jr., and V. McKoy, J. Phys. B 7, 2396 (1974).
23. H. P. Kelly, Phys. Rev. 136, B896 (1964).
24. D. H. Madison and W. N. Shelton, Phys. Rev. A 7, 499 (1973).
25. G. Csanak, J. Phys. B 7, L203 (1974).
26. R. R. Lucchese, D. K. Watson, and V. McKoy, Phys. Rev. A 22, 421 (1980).
27. S. Chung and C. C. Lin, Phys. Rev. A 17, 1874 (1978).
28. C. A. Weatherford, Phys. Rev. A 22, 2519 (1980).
29. T. K. Holley, S. Chung, C. C. Lin, and E. T. P. Lee, Phys. Rev. A 24, 2946 (1981).
30. K. Takatsuka and V. McKoy, Phys. Rev. A 24, 2473 (1981).
31. K. Takatsuka and V. McKoy, Phys. Rev. A (to be published).
32. Marco A. P. Lima, Thomas L. Gibson, K. Takatsuka, and V. McKoy, Phys. Rev. A (to be published).
33. S. Trajmar, D. C. Cartwright, J. K. Rice, R. T. Brinkmann, and A. Kuppermann, J. Chem. Phys. 49, 5464 (1968).
34. S. J. B. Corrigan, J. Chem. Phys. 43, 4381 (1965).
35. S. K. Srivastava and S. Jensen, J. Phys. B 10, 3341 (1977).
36. D. C. Cartwright, A. Chutjian, S. Trajmar, and W. Williams, Phys. Rev. A 16, 1013 (1977).
37. A. Chutjian, D. C. Cartwright, and S. Traymar, Phys. Rev. A 16, 1052 (1977).

HIGH ENERGY ELECTRON SCATTERING AND ELECTRON CORRELATION

IN ATOMS AND MOLECULES

R. A. Bonham
Chemistry Department
Indiana University
Bloomington, Indiana 47405

(Abstract)

Recent measurements of the one and two electron potential energies in Ne show conflicting results for the partitioning of correlation effects. The need for further measurements and theoretical calculations is stressed.

In the case of molecules the one and two electron contributions to the chemical binding energy are found in certain cases to be extremely sensitive to electron correlation. A number of cases are found where Hartree-Fock theory gives a totally incorrect picture of one and two electron contributions to chemical binding although the estimated total binding energy is reasonable.

Recently Duguet (1) has measured by means of high energy electron scattering both the x-ray coherent, $F(K)$, and x-ray incoherent, $S(K)$, scattering factors as a function of momentum transfer for atomic Ne and Ar. The discussion here is limited to the case of Ne since no adequate theory or alternative experimental measurements yet exist for Ar.

A knowledge of $F(K)$ over a sufficient range in K permits one to calculate the electron nuclear attractive energy, $\bar{V}_{en}$, in atomic units as (2)

$$\bar{V}_{en} = - \frac{2Z}{\pi} \int_0^\infty dK\, F(K) \tag{1}$$

and the classical or Coulomb electron-electron repulsive energy as (3)

$$\bar{V}_{ee}{}^c = \frac{2}{\pi} \int_0^\infty dK\, F(K)^2 \tag{2}$$

On the other hand a knowledge of the incohreent scattering factor over an extended range in K makes it possible to define the difference between the electron-electron repulsive potential energy, $\bar{V}_{ee}$, and $\bar{V}_{ee}{}^c$ as

$$\bar{V}_{ee} - \bar{V}_{ee}{}^c = \frac{2}{\pi} \int_0^\infty dK\, [S(K)-N] \tag{3}$$

where N denotes the number of electrons. Since $\bar{V}_{ee}{}^c$ is known from Eq. 2 it is possible in principle to determine both $\bar{V}_{en}$ and $\bar{V}_{ee}$ for atoms by use of high energy electron scattering or x-ray scattering. In the case of Ne Duguet considered the difference functions $F(K)_{expt} - F(K)_{HF}$ and $S(K)_{expt} - S(K)_{HF}$ where the subscripts refer to the experiment and Hartree-Fock theory respectively. The energies derived from these difference functions are then the correlation energy contributions to $\bar{V}_{en}$ and $\bar{V}_{ee}$. In Table I the results are given for Duguet's measurements along with theoretical results from a CI wave function yielding 86% of the correlation energy (4) and a perturbation (Bethe-Goldstone) wave function yielding 100% of the correlation energy (5). Also included are experimental estimates based on known ionization potentials (6). Because the theoretical values are derived from calculated values of S(K) and F(K) which were not computed with sufficient accuracy the quoted theory results are uncertain by as much as ± 2 eV. The qualitative picture is that the Duguet and perturbation results agree with each other but differ from the CI and ionization potential results. The results are summarized in Table I. While the uncertainties on the potential energies are large enough to call into question the significance of the observed deviations the actual differences in the shape of $\Delta F(K)$ are striking. Recently Fink et al (7) have repeated the measurements of Duguet and found good agreement with his results if the same range in momentum transfer was used. However when the experiment was carried out to larger K values the picture changed completely and the results came close to those for the CI calculation and ionization potentials. What remains to be done is for someone to undertake direct computation of ΔV_{ee} and ΔV_{en} using a highly correlated wave function. Obviously experiments with increased precision would also be desirable. The major conclusion so far is that measurements of $\Delta F(K)$ are obviously very sensitive to the electron correlation and can serve as a useful diagnostic of wave function quality even when the derived energy is rather uncertain.

Table I. Correlation effects on $\bar{V}_{en}$, $\bar{V}_{ee}$ and the total electronic energy for Ne in electron volts

	Configuration Interaction (CI) Ref. 4 [*]	Perturbation Theory Ref. 5 [*]	Experiment	
			Electron Scattering Ref. 1	Ionization Potentials Ref. 6
$\Delta\bar{V}_{en}$	+ 3.8 ± 2.0	- 6.2 ± 2.0	- 7.3 ± 7.0	+ 1.4 ± 0.7
$\Delta\bar{V}_{ee}$	-17.2 ± 2.0	-18.6	- 7.6 ± 7.0	-22.3 ± 0.7
E_c	- 6.7 ± 2.0	-12.4 ± 2.0	- 9 ± 5.0	-10.5 ± 0.4

[*] The uncertainties in these columns are estimated on the basis of numerical integration of form factor data interpolated from graphs.

In the case of molecules the type of analysis outlined above for atoms is no longer possible since one cannot extract F(K) directly from experimental data (8). It is possible however to extract the potential energy quantities $\bar{V}_{en} + \bar{V}_{ee}^c$ from the elastic scattering and $\bar{V}_{ee} - \bar{V}_{ee}^c$ from the inelastic scattering. The physical meaning of these two terms is clear. The elastic scattering yields the total potential energy of the molecule accessible from a knowledge of the exact diagonal one electron density matrix while the inelastic part yields the remainder. One would normally expect the latter to be more sensitive to electron correlation effects. While only a few experimental results are available a number of theoretical calculations are available at the Hartree-Fock level. In order to view these results in a sensitive manner we may calculate the elastic and inelastic scattering for all the separate atoms in a molecule using the best available atomic wave functions. One can then subtract the sum of the atomic intensities, each weighted by the number of times each particular atom type occurs in the molecule, from the experimental intensities. Such a difference intensity as a function of momentum transfer is usually referred to as a $\Delta\sigma(K)$ function with a subscript el or inel denoting the elastic or inelastic contribution. The area under a $\Delta\sigma(K)$ curve is then proportional to the binding potential energy contribution from either the one electron or two electron terms as defined previously.

In Table II the results for the various energies from Hartree-Fock calculations of the elastic and inelastic $\Delta\sigma(K)$ functions are shown. The atomic comparison model is based on Hartree-Fock atoms. The calculations on N_2, O_2 and CO were carried out by Epstein and Stewart (9) and for ethane by Pulay (10). These results are quite striking since in 3 out of the 4 cases the two electron contribution acts to destabilize the chemical bond(s). Only in the case of CO does the two electron term make a substantial contribution to the chemical binding. What actually happens in the real world? Unfortunately until very recently we had no idea. Several years ago Johnson et al (11) measured the inelastic scattering of ethane. From these measurements we can infer that the two electron term may contribute as much as 40% of the total binding energy. Just in the past months Schweig and coworkers (12) have shown by using a highly correlated wave function, that in the case of N_2, electron correlation makes a major change in the two electron potential energy. In addition they also found important changes in the one electron potential energy terms.

Table II. Contributions to the binding energy for various parts of the potential energy for selected molecules in the Hartree-Fock approximation

Molecule	N_2 (eV) Ref. 9	O_2 (eV) Ref. 9	CO (eV) Ref. 9	Ethane (eV) Ref. 10
One Electron Potential Energy $V_{ee} - V_{ee}^c$	- 14.6	- 9.63	- 7.50	- 50.0
Two Electron Potential Energy $V_{ee} - V_{ee}^c$	+ 4.06	+ 2.32	- 8.10	+ 2.0
Total Binding Energy (H.F.)	- 5.27	- 3.65	- 7.80	- 24.0
Experimental Total Binding Energy	- 9.76	- 5.12	- 11.24	- 29.33

While there are still too few examples from which to draw any conclusions the picture which is emerging so far appears to suggest that nature is capricous. That is, we have three examples at the HF level where two electron contributions appear to be antibond forming, and one example at the HF level where two electron terms make definite contributions to binding. On the other hand CI theory and experiment indicate substantial binding contributions from two electron terms in all those cases so far studied. In the case of N_2 the CI calculations indicate non trivial correlation contributions even to the one electron binding terms!

It needs to be emphasized that while scattering experiments can yield information about potential energy contributions to the chemical binding its real value will ultimately stem from determination of the shape of the $\Delta\sigma(K)$ functions which provide a much more detailed picture of binding alterations in density functions, albeit in momentum space, than a single value of an energy. It also needs to be stressed that analysis of chemical binding or electron correlation in terms of potential energy components is not unique. An equivalent analysis is possible in principal by use of the kinetic energy. Such an analysis could be carried out by means of x-ray, γ-ray or electron scattering measurement of the Compton profile.

References

1. A. Duguet, These de docteur d'etat, Universite de Paris-Sud, Centre d'Orsay (Juin, 1981).
2. J. N. Silverman and Y. Obata, J. Chem. Phys. $\underline{38}$, 1254 (1963).
3. C. Tavard, M. Rouault and M. Roux, J. Chim. Phys. $\underline{62}$, 1410 (1968).
4. E. M. A. Peixoto, C. F. Bunge and R. A. Bonham, Phys. Rev. $\underline{181}$, 322 (1968).
5. M. Naon and M. Cornille, J. Phys. $\underline{B4}$, 1210 (1971); $\underline{5}$, 1965 (1972).
6. R. R. Goruganthu and R. A. Bonham, Phys. Rev. $\underline{A26}$, 1 (1982).
7. M. Fink and J. J. McCelland, private communication, 1983; [see accompanying article in this volume].
8. R. A. Bonham and M. Fink, "High Energy Electron Scattering", (Van Nostrand-Reinhold Co., New York, 1974).
9. J. Epstein and R. F. Stewart, J. Chem. Phys. $\underline{66}$, 4057 (1977).
10. P. Pulay, private communication, 1982.
11. J. Johnson and G. G. B. de Souza, private communication, 1981.
12. M. Breitenstein, A. Endesfelder, H. Meyer, A. Schweig and W. Zittlau, Chem. Phys. Lett. $\underline{97}$, 403 (1983).

LINE-SHAPE ANALYSIS OF ENERGY-LOSS SPECTRA AND PHOTOELECTRON SPECTRA

Isao Shimamura

RIKEN (The Institute of Physical and Chemical Research)
Wako, Saitama 351, Japan

I. INTRODUCTION

By analyzing energy-loss spectra of electrons or ions scattered by molecules, it is possible, in principle, to determine cross sections for excitation or deexcitation of the molecules in collisions. The cross sections for rotational transitions, however, are difficult to obtain in this way, because the energy resolution of spectrometers is usually insufficient for resolving the closely spaced rotational-transition peaks in the energy-loss spectra.

Take, for example, a homonuclear diatomic molecule in a $^1\Sigma$ state that may be regarded as a linear rotator. The rotational energy depends on the rotational quantum number J as

$$E(J) = BJ(J+1). \tag{1}$$

The rotational constant B is, for example, 7.4 meV for H_2, 0.25 meV for N_2, and 0.019 meV for Na_2. Because of the small B, and hence, of the small rotational level spacing, molecules (except for H_2) in a gas in equilibrium at room temperature are distributed among many different rotational states J. Therefore, an energy-loss spectrum contains many peaks (or rotational lines) corresponding to different rotational transitions $J \rightarrow$.

When a molecule makes a transition $J \rightarrow J' = J + \Delta J$, a scattered electron loses its energy by

$$\Delta E = B[J'(J'+1) - J(J+1)] = 2B\Delta J[J + (\Delta J + 1)/2]. \tag{2}$$

Only even values of ΔJ are allowed for homonuclear diatomics. All rotationally elastic peaks $J \rightarrow J$ ($\Delta J = 0$) are located at the same ΔE (=0). The rotational lines for the $\Delta J = \pm 2$ transitions are spaced at the intervals of 4B, which is about 1 meV for N_2 and about 0.08 meV for Na_2. The $\Delta J = \pm 4$ lines are spaced by 8B. Energy-loss experiments with the energy resolution better than ~ 10 meV are extremely difficult. Therefore, except for the hydrogen, the rotational lines are usually unresolved, and a broad feature containing many rotational lines is observed.

All rotational lines have practically the same shape, called the apparatus function, determined by the spectrometer. Each line is shifted from the elastic line by the energy loss ΔE of Eq.(2), and has an intensity proportional to the cross section $d\sigma(J \rightarrow J')/d\omega$ and to the number N(J) of molecules in state J. Therefore, the energy spectrum of the scattered electrons may be written as the sum

$$I(E') \propto \sum_{J,J'} N(J) [\text{ } \bigwedge \text{ shifted by } \Delta E] \frac{d\sigma}{d\omega}(J \rightarrow J') \tag{3}$$

of contributions from all transitions belonging to a particular rotational band. Here, $\bigwedge$ denotes the apparatus function. Because so many unknown quantities $d\sigma(J \rightarrow J')/d\omega$ contribute to a single broad feature, Eq.(3) as it is is impossible to use for the purpose of deconvolution of the spectrum. The cross sections $d\sigma(J \rightarrow J')/d\omega$ for all values of J and J', however, are expressible in terms only of those for J=0. Furthermore, only a few of the transitions from J=0 occur appreciably in electron-molecule collisions. Hence, Eq.(3) contains practically only a few unknown parameters $d\sigma(0 \rightarrow J'')/d\omega$, which may be determined by fitting Eq.(3) to an observed spectrum.[1] This shall be explained in Sec. III below.

II. DECOMPOSITION INTO ROTATIONAL (ΔJ) BRANCHES

In a gas in equilibrium at room temperature molecules other than the hydrogen are distributed mostly in high-J states. Since the cross sections $d\sigma(J \to J'=J \pm \Delta J)/d\omega$ for a fixed ΔJ approach a finite limiting value $d\sigma_\infty(|\Delta J|)/d\omega$ as $J \to \infty$, the cross sections in Eq.(3) may be approximated by this high-J limit.[2,3] Alternatively, one may multiply the high-J limit by a statistical weight to ensure the detailed balance, and may write[4]

$$\frac{d\sigma}{d\omega}(J \to J'=J+\Delta J) \cong (\frac{2J'+1}{2J+1})^s \frac{d\sigma_\infty}{d\omega}(|\Delta J|) \qquad \text{(for high J)}, \qquad (4)$$

where $s=\frac{1}{2}$ for a linear rotator and $s=1$ for a spherical-top rotator; note that Eqs.(1), (2), and (3) are valid also for spherical-top rotators.

Substitution of Eq.(4) into Eq.(3) leads to a decomposition[2-4]

$$I(E') \underset{\sim}{\propto} \sum_{\Delta J} \tilde{I}_{\Delta J}(E') \frac{d\sigma_\infty}{d\omega}(|\Delta J|) \qquad (5)$$

of the energy-loss spectrum into branches $\tilde{I}_{\Delta J}(E')$, each of which is characterized by the change ΔJ in the rotational quantum number and is defined by

$$\tilde{I}_{\Delta J}(E') = \sum_J N(J) [\bigwedge \text{ shifted by } \Delta E] (1+\frac{2\Delta J}{2J+1})^s . \qquad (6)$$

It is possible to measure the apparatus function $\bigwedge$, or the elastic peak, by replacing the molecular gas by an atomic gas; for atoms the elastic peak is easily resolved from inelastic peaks because of the absence of rotational and vibrational degrees of freedom. One may usually assume the Boltzmann distribution for N(J). Then, the branch spectra $\tilde{I}_{\Delta J}(E')$ are known, and the only unknown quantities $d\sigma_\infty(|\Delta J|)/d\omega$ in Eq.(5) may be determined by least-squares fit of Eq.(5) to any observed spectrum;[2-4] cross sections $d\sigma_\infty(|\Delta J|)/d\omega$ for only a few small values of $|\Delta J|$ are appreciable. Jung et al.[4] apply this method to the e+N_2 collisions and calculate average ΔJ-transition cross sections

$$\frac{d\bar{\sigma}}{d\omega}(\Delta J;T) = \sum_J N(J) (1+\frac{2\Delta J}{2J+1})^s \frac{d\sigma_\infty}{d\omega}(|\Delta J|), \qquad (7)$$

which depend on the gas temperature T because of the dependence of N(J) on T.

III. DECOMPOSITION INTO ANGULAR-MOMENTUM-TRANSFER (J_t) BRANCHES

Under most experimental conditions the rotational periods of the molecules are much longer than the collision time, and the rotational sudden approximation is valid for the cross sections for electron-molecule collisions.[1] In this approximation the cross section $d\sigma(J \to J')/d\omega$ is the sum of contributions from different values of the angular-momentum transfer J_t from the electron to the molecule, $\vec{J_t}$ being $\vec{J'}-\vec{J}$ (see Fig. 1). Each contribution is proportional to the probability $P(J_t;J,J')$ of the angular-momentum transfer J_t when J changes into J', and to the cross section $d\sigma_{AMT}(J_t)/d\omega$ for the transfer J_t of the angular momentum. The cross sections $d\sigma_{AMT}(J_t)/d\omega$ in the sudden approximation are independent of the rotational motion before and after the collision, and hence, are equal to $d\sigma(0 \to J_t)/d\omega$,

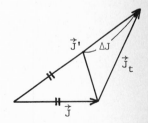

Fig.1. Angular-momentum relation

because $\vec{J}'=\vec{J}_t$ if $\vec{J}=0$. Thus, we have

$$\frac{d\sigma}{d\omega}(J\rightarrow J') = \sum_{J_t} P(J_t;J,J') \frac{d\sigma}{d\omega}(0\rightarrow J_t).$$ (8)

The probability $P(J_t;J,J')$ is given by[5,6]

$$P(J_t;J,J') = \{C(JJ_tJ';000)\}^2$$ (9a)

for linear rotators in terms of a Clebsch-Gordan coefficient, and by[1,5,7]

$$P(J_t;J,J') = \frac{(2J'+1)}{(2J+1)(2J_t+1)}$$ (9b)

for spherical-top rotators.

Equation (8) used in Eq.(3) leads to a decomposition[1]

$$I(E') \underset{\sim}{\propto} \sum_{J_t} I_{J_t}(E') \frac{d\sigma}{d\omega}(0\rightarrow J_t)$$ (10)

of the energy-loss spectrum into angular-momentum-transfer branches defined by

$$I_{J_t}(E') = \sum_{J,J'} N(J)[\,\bigwedge\ \text{shifted by }\Delta E]\,P(J_t;J,J').$$ (11)

Since the branch spectra (11) are known, and since the cross sections $d\sigma(0\rightarrow J_t)/d\omega$ for only a few values of J_t are appreciable, these cross sections can be determined by least-squares fit of Eq.(10) to any measured spectrum.[1] The cross sections for transitions from states other than the ground rotational state are calculable from $d\sigma(0\rightarrow J_t)/d\omega$ by use of Eq.(8).[1] This procedure has been applied successfully to scattering of electrons by nitrogen molecules.[8]

IV. COMPARISON BETWEEN THE TWO METHODS

By use of the decomposition into angular-momentum-transfer branches it is possible to extract from observed energy-loss spectra cross sections $d\sigma(J\rightarrow J')/d\omega$ for transitions from all states J including low-lying states. On the other hand the high-J approximation essential to the decomposition into ΔJ branches precludes the possibility of determining $d\sigma(J\rightarrow J')/d\omega$ for low J by use of the method of Sec.II.

High-J approximations may be made also in the method of angular-momentum-transfer branches. Expansion of the Clebsch-Gordan coefficient in Eq.(9a) in inverse powers of $(2J+1)$ gives an approximate formula[*,1,5,9]

$$P(J_t;J,J+\Delta J) \cong A_{|\Delta J|,J_t}(1+\frac{2\Delta J}{2J+1})^s \qquad \text{(for high J)},$$ (12)

which is correct within an error of the order of $(2J+1)^{-2}$. The matrix $\underset{\sim}{A}$ depends only on $|\Delta J|$ and J_t and not on J:

$$A_{|\Delta J|,J_t} = (2J_+)!(2J_-)!/[2^{J_t}(J_+!)(J_-!)]^2 \qquad (2J_{\pm}=J_t\pm\Delta J)$$ (12a)

---

* Reference 1 gives an equivalent expansion of $P(J_t;J,J+\Delta J)$ in inverse powers of J. This formula has been used later by Chang[10] for essentially the same purpose of high-J approximation as in Ref.1.

for linear rotators for $J_t = |\Delta J|$, $|\Delta J| + 2$, $|\Delta J| + 4$, $\cdots$, and otherwise $A_{|\Delta J|, J_t} = 0$. Equation (12) is exact for spherical-top rotators with

$$A_{|\Delta J|, J_t} = (2J_t + 1)^{-1} \tag{12b}$$

for $J_t = |\Delta J|$, $|\Delta J| + 1$, $|\Delta J| + 2$, $\cdots$ and $A_{|\Delta J|, J_t} = 0$ for $J_t < |\Delta J|$. Equation (12) used in Eq. (11) gives an approximation to the spectra $I_{J_t}(E')$ in the decomposition formula (10). Although this is a high-J approximation, it still allows one to determine the cross sections $d\sigma(0 \rightarrow J_t)/d\omega$, and hence, $d\sigma(J \rightarrow J')/d\omega$ for all J from Eq. (8), just as is described in Sec. III.

This high-J approximation may be related to that of Sec. II. The use of Eqs. (11) and (12) in Eq. (10) results in a decomposition

$$I(E') \underset{\sim}{\propto} \sum_{\Delta J} \tilde{I}_{\Delta J}(E') [\sum_{J_t \geq |\Delta J|} A_{|\Delta J|, J_t} \frac{d\sigma}{d\omega}(0 \rightarrow J_t)]. \tag{13}$$

Comparison between Eqs. (5) and (13) reveals a relation

$$\frac{d\sigma_\infty}{d\omega}(|\Delta J|) = \sum_{J_t \geq |\Delta J|} A_{|\Delta J|, J_t} \frac{d\sigma}{d\omega}(0 \rightarrow J_t). \tag{14}$$

If Eq. (14) is solved for $d\sigma(0 \rightarrow J_t)/d\omega$ and Eq. (8) is used, the cross sections $d\sigma(J \rightarrow J + \Delta J)/d\omega$ for all $J \geq 0$ are calculable from the cross sections $d\sigma_\infty(|\Delta J|)/d\omega$ extracted from experiment by the method of Sec. II. Figure 2 shows a result of this procedure applied to those integral cross sections $\sigma_\infty(|\Delta J|)$ for $e+N_2$ collisions which Jung et al.[4] have obtained by analyzing the energy-loss spectra they have measured. Theoretical cross sections due to Chandra and Temkin[11] are also included in the figure for comparison.

The number of fitting parameters to be used in the J_t-branch method is sometimes less than that in the ΔJ-branch method due to the symmetry property of the molecules. This is because the J_t branches are physically more meaningful than the ΔJ branches that have been introduced from a technical viewpoint. For the spherical-top molecule CH_4, for example, the values 1 and 2 of the angular-momentum transfer J_t are forbidden, whereas all values of ΔJ are allowed in general for high J. Suppose that the cross sections $d\sigma(0 \rightarrow J_t)/d\omega$ for $J_t \geq 5$ are negligibly small. Then the fitting parameters in the J_t-branch method are $d\sigma(0 \rightarrow J_t)/d\omega$ for $J_t = 0$, 3, and 4. To attain the same accuracy in the ΔJ-branch method one has to use five fitting parameters $d\sigma_\infty(|\Delta J|)/d\omega$ for $|\Delta J| = 0$, 1, 2, 3, and 4. In fact, only three of these five parameters are linearly independent theoretically, although this is not apparent in the conventional derivation of the ΔJ-branch method.

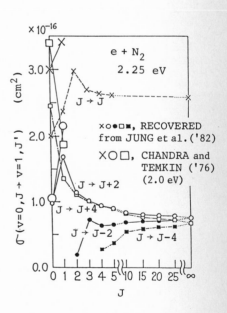

Fig. 2. Experimental and theoretical integral cross sections for vibrational-rotational transitions $(v, J) = (0, J) \rightarrow (1, J')$ in $e+N_2$ collisions at the highest maximum in the $^2\Pi_g$ resonance region. See text for explanation.

V. DISCUSSION

The least-squares fit produces meaningful results on rotationally inelastic collisions only when the rotational broadening compared with the apparatus function is clearly observed in the energy-loss spectrum. The rotational broadening is due to th terms in Eq. (10) other than the $J_t=0$ term. For the effect of the terms with $J_t \neq 0$ to be observable, the following conditions must be satisfied:

(1) The ratios $[d\sigma(0 \to J_t)/d\omega]/[d\sigma(0 \to 0)/d\omega]$ for $J_t \neq 0$ are appreciable.
(2) The energy loss by electrons is so large as to make $I_{J_t}(E')$ for $J_t \neq 0$ much broader than the apparatus function $I_{J_t=0}(E')$. This requires either B to be large (the molecule to be light) or the rotational temperature T of the molecular gas to be high. Applications to molecules much heavier than N_2 are difficult with the apparatus resolution of ~ 10 meV.
(3) The energy resolution of the spectrometer is high, and the energy-loss spectrum has a tail sharply cut off on either side of the broad feature.
(4) The statistics of the data is good.

The angular-momentum-transfer branch method is easily generalizable for symmetri top rotators. They have a well-defined quantum number K for the projection of the ro tational angular momentum onto the body-fixed symmetry axis. The cross section for a transition $JK \to J'K'$ is expressible as

$$\frac{d\sigma}{d\omega}(JK \to J'K') = \sum_{J_t} \{C(JJ_tJ';KK_tK')\}^2 \frac{d\sigma}{d\omega}(00 \to J_tK_t), \tag{15}$$

which may be used in place of Eq. (8) in Sec. III. High-J formulas are also available

A similar procedure may be used, in principle, for asymmetric-top rotators,[13] bu the number of fitting parameters is larger in this case and the procedure may be impractical.

So far only rotational quantum numbers have been indicated explicitly for specif ing the molecular transition. The procedures of this paper, however, are equally appl cable to simultaneous rotational-vibrational transitions, provided that the rotationa structure of a particular vibrational transition be studied. The result shown in Fig is, in fact, an example of such a case.

The methods described in this paper are also applicable to photoelectron spectra Eqs. (5), (8), (10), and (14) are valid for photoionization of molecules M, $h\nu + M(J) \to M^+(J') + e$, and Eq. (15) is valid for photoionization $h\nu + M(JK) \to M^+(J'K') + e$, provided that the rotational sudden approximation be good for the final continuum state.

REFERENCES

1) I. Shimamura: Chem. Phys. Letters 73, 328 (1980).
2) F. H. Read: J. Phys. B5, 255 (1972).
3) S. F. Wong and L. Dubé: Phys. Rev. A17, 570 (1978).
4) K. Jung, Th. Antoni, R. Müller, K.-H. Kochem, and H. Ehrhardt: J. Phys. B15, 3535 (1982).
5) I. Shimamura: Chap. II in Electron-Molecule Collisions, eds. I. Shimamura and K. Takayanagi (Plenum, New York, 1984).
6) S. I. Drozdov: Sov. Phys. JETP 3, 759 (1956).
7) I. Shimamura: J. Phys. B15, 93 (1982).
8) H. Tanaka, L. Boesten, and I. Shimamura: VIIth International Conference on Atomic Physics, MIT, Massachusetts, Aug. 1980, Book of Abstracts, p.43.
9) I. Shimamura: Phys. Rev. A28, 1357 (1983).
10) E. S. Chang: J. Phys. B15, L873 (1982).
11) N. Chandra and A. Temkin: Phys. Rev. A14, 507 (1976).
12) I. Shimamura and A. C. Roy: in preparation.
13) D. W. Norcross: Phys. Rev. A25, 764 (1982).

HOLE LOCALIZATION EFFECTS IN MOLECULAR INNER SHELL EXCITATION

T. N. Rescigno and A. E. Orel

Theoretical Atomic and Molecular Physics Group
Lawrence Livermore National Laboratory
University of California
Livermore, California 94566

Introduction

Ab initio calculations of valence shell ionization potentials have shown that orbital relaxation and correlation differences usually make contributions of comparable magnitude. In marked contrast to this observation is the situation for deep core ionization, where correlation differences (~1 eV) play a relatively minor role compared to orbital relaxation (~20 eV). Theoretical calculations have shown that this relaxation is most easily described if the 1s-vacancy created by a K-shell excitation is allowed to localize on one of the atomic centers.[1] For molecules possessing a center of inversion, this means that the molecular orbitals that best describe the final state do not transform as any irreducible representation of the molecular point group.

This point has been thoroughly investigated for the case of 1s→π* excitation in molecular nitrogen.[2] In this previous study, we found that the $1s^{-1}\pi*$ excited states could be adequately described at the level of single-excitation configuration-interaction using broken symmetry orbitals, the dominant effect being the relaxation of the outer valence orbitals in the presence of a localized 1s-hole. The calculations gave two quasi-degenerate molecular states of $^1\pi_g$ and $^1\pi_u$ symmetry; we also obtained the generalized oscillator strengths[3] for these transitions as a function of momentum transfer.

Recent experimental work by Shaw, King, Read and Cvejanovic[4] and by Stefani and coworkers[5] has prompted us to carry out further calculations on N_2, as well as analogous investigations of $1s_N$→π* excitation in NO and N_2O. The generalized oscillator strengths display a striking similarity and point to the essential correctness of the localized hole picture for N_2. The theoretical calculations are briefly described in the next section, followed by a summary of the results and comparison to experiment, followed by a short discussion.

Theoretical Calculations

The wavefunctions used in this study were all based on limited configuration interaction expansions with molecular orbitals obtained from SCF calculations on the ground state of the molecule. The N_2 calculations have been described in detail previously.[2] The basis sets chosen for the N_2, NO and N_2O calculations were all based on a (5s/3p) contraction of a (9s/5p) set of primitive gaussian functions[6]. These were augmented with two d_π polarization functions as well as one additional diffuse p-function. For the N_2 calculations, "localized" 1σ orbitals are constructed by taking

$$1\sigma_{L,R} = 1\sigma_g \pm 1\sigma_u.$$

The calculations are then carried out in a reduced symmetry ($C_{\infty v}$). Because the 1σ orbitals have a very small overlap with the remaining molecular orbitals, the latter can still be classified by g or u symmetry, as can the total molecular wavefunctions, when a balanced set of reference configurations are used. Starting with the four reference configurations,

$$(1\sigma_R^2 1\sigma_L 2\sigma_g^2 2\sigma_u^2 3\sigma_g^2 1\pi_u^4 2\pi_g), \quad {}^{1,3}\Pi$$

$$(1\sigma_R^2 1\sigma_L 2\sigma_g^2 2\sigma_u^2 3\sigma_g^2 1\pi_u^4 2\pi_u), \quad {}^{1,3}\Pi$$

$$(1\sigma_R 1\sigma_L^2 2\sigma_g^2 2\sigma_u^2 3\sigma_g^2 1\pi_u^4 2\pi_g), \quad {}^{1,3}\Pi$$

$$(1\sigma_R 1\sigma_L^2 2\sigma_g^2 2\sigma_u^2 3\sigma_g^2 1\pi_u^4 2\pi_u), \quad {}^{1,3}\Pi$$

we include all single excitations relative to each configuration. Note that the orbitals we have labelled $2\pi_{g,u}$, which are unoccupied in the N_2 ground state, are chosen by inspecting the virtual orbitals and selecting those which have the largest spatial overlap with the atomic nitrogen 2p.

The calculations on NO and N_2O were similarly executed. The distinguishability of the nuclear centers, of course, obviates the need for explicit construction of localized core orbitals. The N_2O ground state is linear (we used R_{NN} = 2.128 a.u. and R_{NO} = 2.243 a.u.) and has the configuration: $(1\sigma^2 2\sigma^2 3\sigma^2 4\sigma^2 5\sigma^2 6\sigma^2 1\pi^4 2\pi^4), {}^1\Sigma^+$. The $1s_N \rightarrow \pi^*$ singlet states were obtained by including all single excitations relative to the configurations $(1\sigma^2 2\sigma 3\sigma^2 4\sigma^2 5\sigma^2 6\sigma^2 1\pi^4 2\pi^4 3\pi), {}^1\Pi$ and $(1\sigma^2 2\sigma^2 3\sigma 4\sigma^2 5\sigma^2 6\sigma^2 1\pi^4 2\pi^4 3\pi), {}^1\Pi$. As in the case of N_2, the 3π orbital was chosen, after inspection of the virtual orbitals, to be an antibonding $2p_\pi$ orbital. The 2σ and 3σ orbitals correspond to the central and outer nitrogen 1s orbitals, respectively. We label these N_C and N_T.

The NO calculations were carried out at the equilibrium internuclear separation of 2.173 a.u. The ground state configuration is $(1\sigma^2 2\sigma^2 3\sigma^2 4\sigma^2 5\sigma^2 1\pi^4 2\pi)$, $^2\Pi$. The reference configurations for the spin-allowed nitrogen core-excited states were obtained by moving an electron from the 2σ orbital into the partially occupied 2π orbital. The configuration-interaction calculations included all single excitations relative to these reference configurations and were carried out for the nitrogen core-excited states of $^2\Sigma^+$, $^2\Sigma^-$ and $^2\Delta$ symmetry.

The limited CI calculations we have carried out were aimed at treating correlation effects in the initial and final states on an equal basis. The appropriate choice of reference configurations for the excited states and the restriction to single excitation CI incorporates the dominant relaxation effects and gives transition energies of essentially ΔSCF quality which, for these systems, should be accurate to roughly 1 eV. The use of CI wavefunctions employing a common set of orthonormal molecular orbitals also simplifies the calculation of properties such as the generalized oscillator strengths, which were obtained for all the spin-allowed transitions.

Results

Table I summarizes the results of our theoretical calculations on the nitrogen $1s \to \pi^*$ excited states of N_2, N_2O and NO. For N_2, the calculated vertical singlet-triplet energy splittings are .94 and .89 eV for the $^{1,3}\Pi_u$ and $^{1,3}\Pi_g$ states, respectively. This can be compared to the .82 eV difference in energy between the v = 0 levels measured by Shaw, et al[4]. These investigators[4,7], however, see no evidence of separate Π_g and Π_u states. The g-u splittings we obtain are ~.1 eV, which is less than the natural (Auger) decay width of the states and therefore, presumably, unresolvable.

In N_2O, we find two $^1\Pi$ states separated by 4 eV. This splitting comes about from the difference in the binding energies of the 1s electrons residing on two non-equivalent nitrogen centers.[8] The calculated transition energies for these states are in excellent agreement with the experimental values obtained by Bianconi et al.[9] using monochromatized sychrotron radiation. There is also good agreement between the calculated optical oscillator strengths and the experimental value of .11 obtained by Stefani and coworkers[5].

The calculated results for NO provide an interesting example of the equivalent nucleus model of K-shell excitation. According to this model, the effect of a K-shell vacancy is to essentially increase the positive charge of the nuclear core by one.[10] There should then be a correspondence between the core-excited states of NO and the valence states of O_2. The core-excited $^2\Sigma^-$, $^2\Delta$ and $^2\Sigma^+$ states of NO we have calculated have as their analogues the $X^3\Sigma_g^-$, $a^1\Delta_g$ and $b^1\Sigma_g^+$ valence states of O_2. Although the ordering of the calculated NO states is the same as the corresponding O_2 states, the energy differences are slightly different.

Table I. Summary of nitrogen ($1s^{-1}\pi^*$) excited states.

Species	Transition Energy	Oscillator Strength
N_2 $(1s^{-1}1\pi_g)$, $^1\Pi_u$	399.06 (400.86[a])	.236 (.195[b])
$(1s^{-1}1\pi_g)$, $^1\Pi_g$	399.12	---
$(1s^{-1}1\pi_g)$, $^3\Pi_u$	398.12 (400.02[c])	---
$(1s^{-1}1\pi_g)$, $^3\Pi_g$	398.23	---
N_2O $(1s_N^{-1}3\pi)$, $^1\Pi$	400.6 (401.16[d])	.090(.11[e])
$(1s_N^{-1}3\pi)$, $^1\Pi$	404.7 (404.92[d])	.119
NO $(1s_N^{-1}2\pi^2)$, $^2\Sigma^-$	397.7	.061
$(1s_N^{-1}2\pi^2)$, $^2\Delta$	397.8	.038
$(1s_N^{-1}2\pi^2)$, $^2\Sigma^+$	399.0	.019

a $^1\Pi_{g,u}$ unresolved, ref. 7
b R. B. Kay, Ph. E. VanderLeeuw and M. J. VanderWiel, J. Phys. B $\underline{10}$, 2513 (1977)
c $^3\Pi_{g,u}$ unresolved, ref. 4
d ref. 9
e ref. 5

Stefani and coworkers have measured the generalized oscillator strengths for the nitrogen $1s \to \pi^*$ transitions in N_2, NO and N_2O. Figure 1 compares our calculated results with the experimental data for the case of N_2. Note that the data is in excellent agreement with the summed $^1\Pi_u$ and $^1\Pi_g$ GOS values, pointing to the enhancement of the dipole-forbidden transition at values of the momentum transfer greater than 1 a.u. Figure 2 compares the summed GOS results for N_2, NO and N_2O to the experimental values. The shapes of the GOS are indistinguishable in the three molecules.

Discussion

The nitrogen $1s \to \pi^*$ excited states in N_2, NO and N_2O are quite similar and are well described by limited configuration-interaction wavefunctions that include the effect of orbital relaxation following creation of a 1s vacancy. For the case of N_2, localized 1s orbitals must be used to describe this relaxation properly. (Analogous calculations using g-u symmetrized orbitals give transition energies in error by 12-13 eV.[2]) The N_2 molecule with a 1s-hole behaves like a heteropolar molecule, although the overall symmetry of the wavefunctions we calculate correspond to $D_{\infty L}$. Similar considerations are expected to apply to 1s-vacancy production in any molecule containing atoms in equivalent locations.

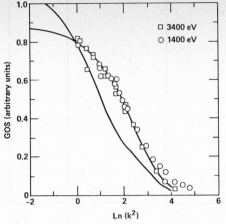

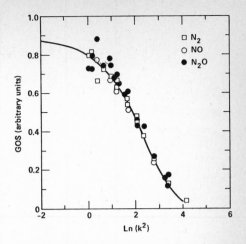

Fig. 1 GOS for 1s→π* excitation in N_2.
Upper curve: summed $X^1\Sigma_g \to ^1\Pi_u$ and $X^1\Sigma_g \to ^1\Pi_g$;
lower curve: $X^1\Sigma_g \to ^1\Pi_u$ only. Experimental
data from reference 5. Theoretical data
arbitrarily normalized to experiment at
k=1.0 a.u.

Fig. 2 Comparison of calculated total
and experimental GOS for 1s→π* excitation
in N2, N_2O and NO. Experimental data from
reference 5.

Acknowledgement

The authors would like to thank Dr. G. Stefani for making his unpublished data
available to them. This work was performed under the auspices of the U. S. Department
of Energy at the the Lawrence Livermore National Laboratory under Contract No.
W-7405-ENG-48.

References

1. L. S. Cederbaum and W. Domcke, J. Chem. Phys., 66, 5048 (1977).
2. T. N. Rescigno and A. E. Orel, J. Chem. Phys., 70, 3390 (1979).
3. M. Inokuti, Rev. Mod. Phys. 43, 297 (1971).
4. D. Shaw, G. King, F. Read and D. Cvejanovic, J. Phys. B, 15, 1785 (1982).
5. G. Stefani (private communication); R. Camilloni, E. Fainelli, G. Petrocelli and
    G. Stefani, International Symposium on Wavefunctions and Mechanisms from Electron
    Scattering Processes, Rome, Villa Montecucco, July23-24, 1983
6. T. H. Dunning, J. Chem. Phys. 53, 2823 (1970).
7. G. C. King, F. H. Read and M. Tronc, Chem. Phys. Letts. 52, 50 (1977).
8. W. H. E. Schwartz, Angew. Chem. Int. Ed. Eng. 13, 454 (1974).
9. A. Bianconi, H. Petersen, F. C. Brown and R. Z. Bachrach, Phys. Rev. A 17, 1907
    (1978).
10. M. Nakamura, M. Sasanuma, S. Sato, M. Watanabe, H. Yamashita, Y. Iguchi, A .
    Ejiri, S. Nakai, S. Yamaguchi, T. Sagawa, Y. Nakai and T. Oshio, Phys. Rev. 178,
    80 (1969).

THE INADEQUACY OF FIRST ORDER THEORETICAL MODELS FOR IONISATION

R.J. Tweed

Faculté des Sciences, Université de Bretagne Occidentale

6, avenue Victor Le Gorgeu, 29283 Brest Cedex, France.

Electron-electron coïncidence experiments for ionisation of atomic targets by electron impact were originally carried out by Ehrhardt et al(1969) in a non-symmetric coplanar geometry. For an incident electron of energy E_0, momentum $\underline{k}_0$ and two final state electrons of energies E_A, E_B and momenta $\underline{k}_A$, $\underline{k}_B$ respectively, the cross section is plotted as a function of $\theta_B = \cos^{-1}(\hat{k}_B.\hat{k}_0)$ for several fixed values of $\theta_A = \cos^{-1}(\hat{k}_A.\hat{k}_0)$. Two lobes generally appear, in directions near to the axis $\underline{K} = \underline{k}_0 - \underline{k}_A$. It is common practice to ascribe the forward "binary" lobe to single interactions between the colliding and ejected electrons, and the backward "recoil" lobe to multiple interactions involving momentum transfer to the residual ion. This interpretation implies that a sufficiently sophisticated first order theoretical model should prove adequate for the binary lobe, but would break down for the recoil lobe.

Following the model of Baluja and Taylor (1976), Madison et al (1977), Bransden et al (1978) and Tweed (1980) carried out similar calculations for electron impact ionisation of helium. Only first order matrix elements of the interaction potential were present ; distorted waves were used to represent the incident, the scattered and the ejected electrons. Exchange terms were included and the convergence of the numerical integrals and partial wave series were carefully checked. The three calculations are in excellent agreement among themselves, in spite of the complete independance of the computer codes used. In figure 1, certain of these theoretical results are compared with experimental data by Pochat et al (1982a) ; Ehrhardt and al (1972). In general, the directions of the lobes move away from the $\hat{K}$ axis, but only by a relatively small angle In the leftmost two plots good agreement is obtained with experiment for the binary lobe ; in the rightmost plot, where the experimental lobe lies considerably away from the $\hat{K}$ axis, theory gives bad results. For the recoil lobe the results of the first order theoretical models are always inadequate.

$E_0 = 100$ eV $_-$ $E_B = 5$ eV $E_0 = 80.5$ eV $_-$ $E_B = 2.5$ eV $E_0 = 80.5$ eV $_-$ $E_B = 15.5$ eV
$E_A = 70.42$ eV $_-$ $\theta_A = 10°$ $E_A = 53.42$ eV $_-$ $\theta_A = 30°$ $E_A = 40.42$ eV $_-$ $\theta_A = 7°$

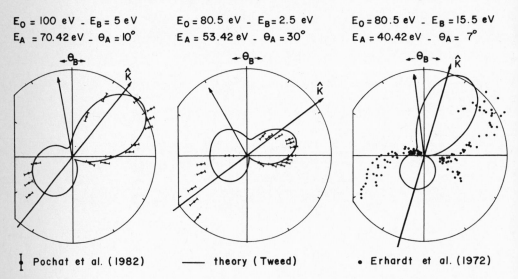

$\frac{1}{1}$ Pochat et al. (1982)   —— theory (Tweed)   • Erhardt et al. (1972)

Figure 1: Comparison of experiment and first order theory for ionisation of helium.

In the experiments of Pochat et al (1982b), also for a helium target, a large range of angles θ_A were considered and it was possible to compare the directions of the binary lobe with that of the $\hat{K}$ axis. The right hand and left hand parts of figure 2 correspond to an inversion of the energies E_A and E_B. In both cases the experimental binary lobe is at a larger angle than $\hat{K}$ for small θ_A and at a smaller angle for large θ_A and will be very near to $\hat{K}$ at $\theta_A \simeq 35°$. This particular behaviour is characteristic of situations where E_A and E_B are of similar magnitude, but we may expect the existance in general of region of "lucky" scattering angles θ_A where the experimental binary lobe is close to the $\hat{K}$ axis. In such regions fortuitously good agreement will probably be found between experiment and sufficiently elaborate first order models.

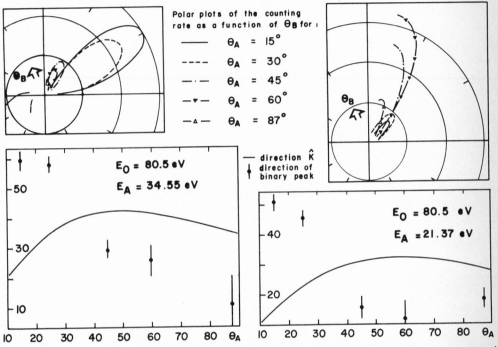

Figure 2 : Comparison of the directions of the experimental binary peak and the $\underline{K}$ axis

The above mentioned theoretical models may be extended to the case of autoionisation following the method of Balashov et al (1972,1973) modified by Tweed (1976) to include exchange. This gives a parametric form for the cross sections ; for (e-2e) experiments it is simpler to compare the angular dependance of resonance parameters,than series of profiles taken at different angles. The wavefunction describing the residual ion and electron B at energies near to that of a particular autoionising state, is locally replaced by a linear combination with a wavefunction representing the latter. Such wavefunctions are of bound state type and the scattering amplitudes for their excitation may be calculated with reasonable accuracy in a first order model. This gives :

$$\frac{d^3\sigma}{d\hat{k}_A\,d\hat{k}_B dE_B} = c(\underline{k}_A,\underline{k}_B) + \sum_\nu\ (2\,S_\nu + 1)\ \frac{a_\nu\ (\underline{k}_A^\nu,\underline{k}_B^\nu)\ \varepsilon_\nu + b_\nu\ (\underline{k}_A^\nu,\underline{k}_B^\nu)}{1+\varepsilon_\nu}$$

where $c(\underline{k}_A,\underline{k}_B)$ is the direct ionisation cross section and ν is an index for the resonances. These have total spins S_ν, decay widths Γ_ν and positions corresponding to energies E_ν of electron B; $\underline{k}_A^\nu,\underline{k}_B^\nu$ are momenta in the directions $\hat{k}_A$, $\hat{k}_B$ respectively, corresponding to these particular energies;the $\varepsilon_\nu=2(E_B-E_\nu)/\Gamma_\nu$ are dimensionless reduced energies. Parameters a_ν and b_ν have the dimensions of cross sections ; their ratio determines the form of the resonance profile and b_ν is proportional to the total contribution of the resonance to the cross section integrated over E_B.

After convolution with the apparatus function this parametric form may be fitted to the experimental data and the parameters plotted as a function of the angle θ_B. The theoretical expressions for the parameters contain cross terms between the amplitudes for direct and auto ionisation ; for the former, it is unlikely that first order models give correct results. However, we may hope that when autoionisation is the dominant process, the resonance profiles will be correctly described, even if the direct ionisation background is incorrect. This situation corresponds to the case of fairly symetric resonances, of relatively large magnitude compared to the background. Calculations by Balashov et al (1980) for a cadmium target give good agreement with the resonance contribution to the triple differential cross section as measured by Martin and Ross (1982) once compensation is made for omitted exchange terms.

For the case of a helium target, experimental a_ν and b_ν parameters have been compared with the results of a theoretical model where exchange terms were included consistently (Pochat et al.1982c). Figure 3 illustrates results for the (2s,2p) ^1P resonance. On the left, triple differential cross sections are compared ; the agreement appears quite satisfactory, especially concerning the presence of multiple lobes. On the right, double differential cross sections (integrated over θ_A) are compared : the agreement is much less good, indicating that the theory breaks down for scattering angles where the direct ionisation process dominates.

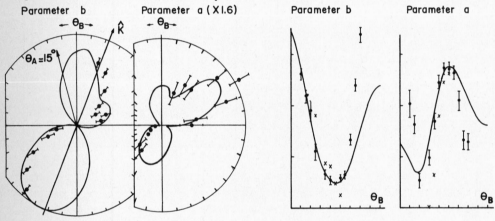

TRIPLE DIFFERENTIAL CROSS SECTION   DOUBLE DIFFERENTIAL CROSS SECTION

Figure 3 : Resonance parameters for the (2s,2p) ^1P autoionising state of helium

Coïncidence experiments for ionisation may also be carried out in a symmetric coplanar geometry. In this case $E_A=E_B=E$ and $\theta_A=\theta_B=\theta$; $\underline{k}_A$ and $\underline{k}_B$ are on opposite sides of $\underline{k}_0$. The variation of the cross section as a function of θ is particularly intersting because the Born I and Born II approximations give completely different angular dependance. After a maximum at around $\theta=20°$ the cross section decreases rapidly with angle ; but whereas in a first-order model it continues to fall as θ increases, in a second order model it passes through a minimum at around $90°$ and then rises sightly to a maximum at about $135°$ before again decreasing. This behaviour, which is evident in the calculations of Byron et al (1982) for a hydrogen target, is in fact independant both of the target and of the details of any particular theoretical model.

Therefore, experiments of this type may be used to provide a rigorous test of whether first order theoretical models for ionisation are adequate, or if it is necessary to go to higher order. Unfortunately, (e - 2e) experiments generally encounter geometrical constraints for $\theta > 100°$ and become progressively more difficult to carry out as the incident electron energy increases, because of the low signal levels at large angle. Such experiments have, however, recently been realised for energies up to 200eV, for electron impact ionisation of helium (Pochat et al. 1983). The results are presented in figure 4. Whereas the points at θ = 45°need accumulation times of only a few hours, those near 120°represent times of 15, 25 and 80 hours, at energies of 100, 150 and 200eV respectively. Each experimental point is the average of at least three such measurements. At 200eV the experimental data is compared with theoretical calculations in Born I and Born II models (Byron et al 1983). Whereas the first order model gives totally wrong behaviour, the results of the second order model are remarkably good. At all three energies, the behaviour of the experimental data is in itself sufficient to eliminate the possibility of using first order theories. A joint article is in preparation.

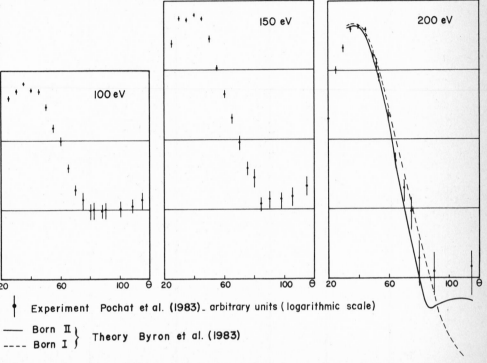

Figure 4 : Triple differential cross sections for He in a symmetric coplanar geometry

Within the range of incident energies 70eV to 300eV (corresponding to the data on which this contribution is based) it is certain that only second or higher order theoretical models can correctly reproduce the results of electron-electron coincidence experiments over an extended range of angles θ_A and θ_B. In the non-symetric coplanar geometry it is possible that at certain scattering angles for the case of direct ionisation, sufficiently elaborate first order models may give reasonable agreement with experiment for the binary peak, provided that the latter does not lie too far away from the $\hat{K}$ axis. They do not in general give correct results for the recoil peak. For the case of autoionisation, a similar type of model is probably capable of giving a correct description of the evolution, as a function of angle, of the forms and magnitudes of the resonance profiles in regions where the autoionisation process dominates direct ionisation. It is possible that the incorporation of second order corrections only in the direct ionisation part of the scattering amplitude would permit the extension of the model to other regions of angle and incident energy.

A rigourous test of any theory (to any order) necessarily implies comparison with experiment over an extended range of scattering angles θ_A, and in particular at angles where the direction of the experimental binary lobe lies well away from the $\hat{K}$ axis. In this context the work of Erhardt et al (1982) is of particular interest. At an incident energy of 500eV the dependance on scattering angle of the direction of the binary peak predicted by a second order model agrees very well with experiment.

The use of a symetric coplanar geometry gives a particularly stringent test of theory, provided that experimental data is available over a sufficiently large range of angles. The experimental data and theoretical calculations currently available confirm the breakdown of first order models at low and medium energies, and seem to indicate that, whereas at 200eV (for helium) a second order model is adequate, at low incident energy, third and higher order terms must also be included. In order to set limits to the energy ranges over which the contributions of terms of any particular order are important, it would be highly desirable to enlarge the domain of incident energies currently covered by theory and experiment for this geometry.

References :

V.V. Balashov, S.S. Lipovetsky and V.S. Senashenko, 1972, Phys. Lett. 40A, 389-90
V.V. Balashov, S.S. Lipovetsky and V.S. Senashenko, 1973, Sov. Phys. JETP 36, 858-60
V.V. Balashov, E.G. Berezhko, A.N. Grum-Grzhimailo, N.M. Kabachnik and A.I. Magunov, 1980, J. Phys. B 13, L269-73
K.L. Baluja and H.S. Taylor, 1976, J. Phys. B 9, 829-35
B.H. Bransden, J.J. Smith and K.H. Winters, 1978, J. Phys. B 11, 3095-114
F.W. Byron Jr., C.J. Joachain and B. Piraux, 1982, 9° Colloque sur la Physique des Collisions Atomiques et Electroniques : Programme et Communications, Nice, FRANCE
F.W. Byron Jr., C.J. Joachain and B. Piraux, 1983, Private Communication
H. Ehrhardt, M. Schulz, T. Tekaat and K. Willmann, 1969, Phys. Rev. Lett. 22, 89
H. Ehrhardt, K.H. Hesselbacher, K. Jung and K. Willmann, 1972, J. Phys. B 5, 1559-71
H. Ehrhardt, M. Fischer, K. Jung, F.W. Byron Jr., C.J. Joachain and B. Piraux, 1982 Phys. Rev. Lett. 48, 1807-10
D.H. Madison, T.C. Calhoun and W.N. Shelton, 1977, Phys. Rev. A 16, 552-62
N.L.S. Martin and K.J. Ross, 1982, J. Phys. B 15, 3959-70
A. Pochat, M. Doritch and J. Peresse, 1982-a, Phys. Lett. 90A, 354-7
A. Pochat, M. Doritch and J. Peresse, 1982-b, 9° Colloque sur la Physique des Collisions Atomiques et Electroniques : Programme et Communications, Nice, FRANCE
A. Pochat, R.J. Tweed, M. Doritch and J. Peresse, 1982-c, J. Phys. B 15, 2269-83
A. Pochat, M. Doritch and J. Peresse, 1983, Private Communication
R.J. Tweed, 1976, J. Phys. B 9, 1725-37
R.J. Tweed, 1980, J. Phys. B 13, 4467-79

(ē,2ē) REACTIONS FOR ATOMS
AND MOLECULES: THE EXPERIMENTS
AND THEORETICAL COMPARISONS

IMPULSIVE (e,2e) EXPERIMENTS: NON DIPOLAR FORM FACTOR*

G. Stefani

Istituto Metodologie Avanzate Inorganiche

C.N.R., Area della Ricerca di Roma

CP10, 00016 Monterotondo, Italy

INTRODUCTION

Aim of this paper is to review relevant results obtained in the field of impulsive (e,2e) spectroscopy over the last few years. (e,2e) experiments, i.e. electron–electron coincidences from impact ionization of atoms and molecules, have recently attracted much attention because of the wealth of information they are providing on both the collisional ionization mechanism and the electronic structure of the target.

The (e,2e) experiment (1,2) consists in measuring simultaneously the energy E_o of the incident electron, the energies E_s and E_e of the two final electrons and the probability that these electron are emitted into solid angles around the directions (ϑ_s, φ_s) and (ϑ_e, φ_e). Depending upon the energy at which the ionization process takes place and the way energy and momenta are shared between the final electrons different theoretical approaches and approximations are used in calculating the triple differential ionization cross-section.

The totality of the interactions leading to ionization events is represented by the continuum of the Bethe surface (3) and by now the (e,2e) experiments have explored almost all the kinematically accesible regions of it. Incident energies ranging between 20 eV (4) and 500 KeV (5) have been used, electronic states as different as the 1S of Silver (5) and the valence orbitals of simple molecules have been explored, atomic systems as simple as H (6) and molecular systems as complex as C_2H_2Br (7) have been investigated.

Therefore, over the last few years, by (e,2e) experiments it has been collected an impressive body of results that are relevant to the ionization mechanism and to the electron spectroscopy of atoms and molecules. Among these experiments this paper will concentrate on the ones for which the binary-encounter model furnishes a realistic description: i.e. the ones belongin to the Bethe-ridge kinematic area. (ē,2e) experiments of different kind are treated in several accompaining papers in this book.

I shall start with studies on the interaction mechanism discussing value and limitation of first order interaction theories in describing high-momentum-transfer ionization events.

Once validity of first order interaction theories is assured the (e,2e) process constitute a spectroscopy of target's electron states and is now being widely applied to the study of several complex atoms and molecules.

VALUE AND LIMITATION OF THE IMPULSE APPROXIMATION

In the Bethe-ridge kinematic region, the angular distribution of the ejected electrons is dominated by a single narrow peak and the cross-section peaks for events where the full amount of energy and momentum exchanged in the collision are directly transferred from the incident electron to a single electron of the target system. It means that the behaviour of the collisional process is dominated by two-body electron-electron interactions.

The more general amplitude M_f for an (e,2e) reaction is written

$$M_f = < \vec{K}_e \ \vec{K}_s (f|T|g > \vec{K}_o > \tag{1}$$

where T is the interaction operator, $\vec{K}_i$ (i=o,e,s) are the incident and outgoing electrons, $|g>$ and $|f>$ are eigenstates respectively of the initial neutral state and final ionic state.

In the Bethe-ridge domain the relation 1 can be greatly simplified by using the binary encounter approximation. It amounts to the assumption that T is a suitable three-body operator. With this approximation (f| commutes with T.

$$M_f = < \vec{K}_e \ \vec{K}_s \ T(f|g > \vec{K}_o > \qquad (2)$$

At high incident energy and high momentum transfer ($\vec{K} = \vec{K}_o - \vec{K}_s$) the op̲erator T can be replaced by a two-body operator and the K 's are well represented by asymptotic free electron wavefunctions. This is the Plane Wave Born Approximation (P W B A); by this approximation the (e,2e) cross-section factorizes exactly as

$$M_f = < \vec{K}' |v_{ee}| \vec{K} > \ < \vec{K}_s \ \vec{K}_e (f|g > \vec{K}_o > \qquad (3)$$

where

$$2\vec{K} = \vec{K}_o + \vec{q}; \quad 2\vec{K}' = \vec{K}_s - \vec{K}_e \ ; \quad \vec{q} = \vec{K}_o - \vec{K}_s - \vec{K}_e$$

v_{ee} is the two-body Coulomb operator.

When the potential operator v_{ee} in relation 3 is replaced by the spin-averaged t-matrix operator the factorization is retained, the collisional term alone changes, and the approximation is called Plane Wave Impulse Approximation (P W I A).

These approximations are particularly helpful in using (e,2e) as a spectroscopic tool because of its factorization in a collision term, due to a two body matrix element, and a form factor which is the Fourier transform of the initial and final states overlap integral.

In recent papers (1,2) it was reviewed that single collision theories are very accurate in describing the "impulsive" triple-differential cross-section. The simple PWIA PWBA have proved to be adequate, provided the incident energy is high and fully symmetric kinematics are used ($\vartheta_s = \vartheta_e$, $E_s = E_e$; $E_o \geq 2$KeV in the case of He). These results stem from extensive experimental work on noble gases done in a variety of symmetric kinematics mainly by the Frascati group (9) and by the Flinders group (2). Validity of these simple frist-order intercation models at high incident energies was also confirmed by meas̲ruements of the absolute value of the (e,2e) cross-section done at F.O.M. on He and H (10) and at Frascati on He, Ne, and Xe (8,11).

By lowering either the incident energy or the degrees of kinematics symmetry of the final state, the (e,2e) cross-section starts showing effects of the interaction of the free electrons between themselves and with the residual ion.

There is no general three-body theory for problems of this kind involving Coulomb potentials; in this kinematic domain (e,2e) constitutes a tool to investigate what to expect from such a theory.

The factorized t-matrix Eikonal distorted Wave Impulse Approximation (EWIA), developped at Flinders, approximates the asymptotic plane waves of the PWIA by plane waves with a wave number shifted to allow for a constant complex distorting potential $\overline{V} + \overline{W}$. In this approximation the factorization is retained: hence its relevance to spectroscopic use of the (e,2e) processes. McCarthy and Weigold (12) have shown that by EWIA all the interactions relevant to a completely symmetric geometry are included, thus introducing in a simple way second order terms in the interaction matrix element (13). Several other computational approaches, as the unfactorized First-Born and Coulomb-Projected-Born, have been used in predicting the impulsive (e,2e) cross-section relative to highly symmetric conditions. None of the approximation used was better than EWIA but the Fully Distorted Impulse Approximation (DWIA)(12), which is computationally very expensive, in accounting for the experimental data (8,9,12). Adequacy of EWIA and PWIA in describing kinematic conditions which are more and more asymmetric has been recently tested in several experiments.

At Flinders experiments on both He and H in asymmetric conditions and intermediate energies have been done (6,14). The experimental findings amount to complete inadequacy of the simpler theories and good agreement over the entire kinematic range explored (incident energy from 800 eV) for the DWIA.

The Frascati group has recently undertaken an experimental investigation on this subject by using He as a target and incident energies between 500 eV and 2000 eV (15). The degree of symmetry of the final state is lowered by reducing both momentum transfer (i.e. the scattering angle) and the kinetic energy of the ejected electron. This investigation clearly shows that the PWIA is a good model for momentum transfer larger than 2 a.u. and for ejected energies larger than 10 eV. Whenever the momentum transfer is reduced below 2 a.u., as shown by the data taken at 2000 eV incident energy and reported in

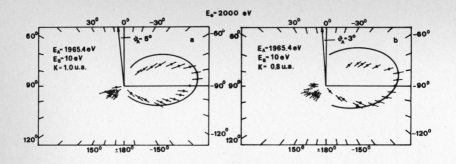

fig. 1 (e,2e) triple differential cross-section for He in asymmetric
        coplanar conditions (Ref.15). Suffixes A and B label the scat-
        tered and ejected electron respectively. Full line is the PWIA.

He 1s Energy sharing
E_0 = 424 eV

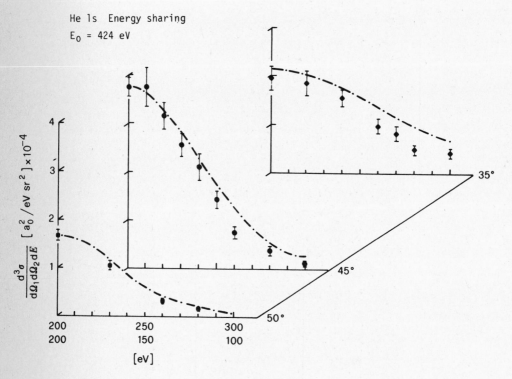

fig 2. (e,2e) triple differential cross-section for He in energy sha-
        ring kinematics measured at 424,5 eV incident energy (Ref.18).
        The full line is the EWIA with V=20 eV and W=0. The double
        scale refers to ejected and scattered electron energies.

fig. 1, the PWIA fails in accounting for the experimental data. Indeed this frist order theory fails in predicting the correct shape for the "binary" lobe of the triple differential cross-section and do not predict at all the "recoil" lobe, which becomes experimentally more and more evident as the momentum transfer is reduced.

Similar experiments, always on He, have been done at even higher incident energy (8KeV) by Lahmam-Bennani et al. (16). These absolute measurements of the triple-differential cross-section show that provided the momentum transfer is larger than about 2 a.u. the simple PWIA and EWIA are adequate in describing the experiment; as they were at lower incident energies. They also show that validity of the EWIA can be extended toward smaller momentum transfer by using a recoil momentum dependent eikonal distorting potential. Distorting potentials of this kind were firstly suggested by Camilloni et al. and applied to their non-coplanar symmetric experiments on the Ne 2p orbital at intermediate energies (17).

The "energy-sharing" is a different experimental approach to the problem of probing the validity of the first order interaction theories when symmetry of the final state is lowered. Experiments of this kind have been started recently on the 1s orbital of He (18,19). They consist in measuring the triple differential cross-section as a function of the recoil momentum $\bar{q}$; this latter quantity being changed by sharing unevenly the kinetic energy between the two final electrons while keeping fixed the scattering and ejection angle ($\vartheta_s = \vartheta_e$). This way the relative momentum ($\bar{K}_e - \bar{K}_s$) is kept almost constant for fixed values of scattering angles and the interaction of the free electron with the residual ion is probed. Results on He at 425 eV (18) are presented in fig. 2. The measured relative cross-section has been brought to the absolute scale by normalizing its value in fully symmetric conditions ($E_s = E_e$; $\vartheta_s = \vartheta_e$) to the value predicted by EWIA, which is in very good agreement with the experimental one measured by van Wingerden et al. (10). The EWIA gives good agreement at the larger angles investigated, but underestimates the cross-section at 35°, thus suggesting that the electron-electron interaction is not well account ed for when the relative energy of the two final electrons becomes small. Popov et al. (20) have treated the Coulomb interactions of the three charged particles in the final state of an (e,2e) process with a semiclassical approach. They obtain, in simple First Born Approximation, good agreement, as far as shapes are concerned, with symmetric coplanar experiments on He and H at 425 eV (21).

To complete this panorama it is to be mentioned that recently (e,2e) experiment on Ag 1s at 500 KeV incident energy have been performed (5) in energy-sharing kinematics. Once more the agreement between experiment and theory improves as the momentum transfer becomes larger.

IMPULSIVE FORM FACTOR AND ELECTRON-CORRELATIONS

As previsouly pointed out symmetric (e,2e) experiments at high incident energy can be interpreted on the basis of the PWIA. Following relation 3, the triple differential cross-section becomes a product of the Mott scattering amplitude for two electrons with appropriate momenta (f_t) by the form factor $\rho(\bar{q})$

$$\frac{d^3 \sigma}{d \Omega_s \, d \Omega_e \, d E} = 4 \, \frac{K_s \, K_e}{K_o} f_t \, \rho(\bar{q}) \tag{4}$$

The collisional factor f_t is either well known or kept almost constant by suitably choosing the kinematic of the experiment (i.e. non-coplanar symmetric conditions). Therefore, under these kinematic conditions, the experiment is a direct measurement of the structure properties of the target through the form factor $\rho(\bar{q})$, which amounts to a squared Fourier Transform with respect to the recoil momentum $\bar{q}$ of the overlap integral (f|g >.

Experiments on the outermost orbitals of noble gases have shown that single particle Hartree-Fock orbitals furnish a satisfactory description for the form factor of closed-shell systems (1,22). Adherence to reality of the Hartree-Fock (H.F.) description has been shown by recent results on the valence orbitals of K and Na metal vapours (23). Sensitivity of the (e,2e) form factor to relativistic effects has been also pointed out by Mitroy and Fuss by calculating the (e,2e) form factor for the 5d orbital of Hg (24).

The (e,2e) spectroscopy is very sensitive to electron-electron correlations when they become a relevant part of the Hamiltonian of the system. This sensitivity is already revealed in the spectroscopy of He. In this case the final ionic state is a one-electron state and any discrepancy with H.F. prediction has to be ascribed to ground state electron-electron correlations. The effect shows up in the ener gy separation spectrum with a "satellite" transition to the n=2 ion states He^+ (2s,2p), at 68 eV separation energy, beside the one to the ground state He^+ (1s) at 24.6 eV. The ratio of intensities of the two transitions as a function of the recoil momentum has been measur ed both by the Flinders group (2) and by the Maryland group (25) and there is excellent agreement between the two experiments.

H.F. calculation predicts transitions only to the (1s) and (2s) states and recoil momentum distribution identical for both transi tions. Experimental results find a rise of this ratio at large re coil momenta, which can be predicted only by using good correlated wave functions or Configuration Interaction description for the ground state of He.

Similar "satellite" transitions have been observed in more com plex atoms. Weigold et al. (26) have carefully studied the (e,2e) transition energy spectrum for the valence shell of Kr and have found a rich "satellite" structure that extendes well above the second ionization continuum threshold. For all the structures above 27 eV transition energy, it was measured a recoil momentum distribution that is characteristic for ionization of the 4s orbital. Therefore the 4s hole strength must be strongly split among a number of ion states, including continuum states, due to strong electron-electron correlations in the final ionic states. Configuration interaction calculation (C.I.) of the Kr separation spectrum extended to the $4s^2 \, 4p^4 \, 5d^1$ configuration misses 27% of the measured pole strength for the 4s electrons ionization and overestimates the fundamental transition strength. The C.I. approach needs to include much more configuration to converge to the experimental value;whether a Green's function (30) approach would be faster in converging is still an open question.

Mixing of sharp transitions and broad transitions to multiply excited states has been observed by Weigold et al. (27) in the vicinities of 4d electrons ionization in Cd. To explain the recoil momentum distribution measured at 18 eV transition energy, that shows a relative maximum at $\bar{q}=0$, (transition to the $4d^9 5s^2 (^1D_{3/2}, ^2D_{5/2})$ ion states) they postulate the presence of two "satellite" transitions ($4d^{10} 6s$ and $4d^{10} 5p$) of the ground state transition ($4d^{10} 5s$).

The occurrence of discrete transition due to single ionization-excitation above the multiple ionization threshold can be fully explained only by abandoning the limitations imposed by a single-particle model. Mixing of sharp transitions and transitions to multiply excited states is shown in fig. 3. This is relative to (e,2e) experiments on Xe performed at 3600 eV incident energy by the Frascati group (28). The reported coincidence spectra are taken in coplanar symmetric geometry at two different scattering angles corresponding to q=0.0 and q=2.0 a.u. The spectra show broad structures, extending from about 23 eV separation energy up to 60 eV, which are assigned to ionization from the 5s orbital plus ecitations on the basis of their common recoil momentum profile. The structure centered at 68 eV has been assigned by photoelectron spectroscopy to the $4d^{-1}$ ion state. Its momentum profile measured by (e,2e) spectroscopy is in agreement with the 4d H.F. orbital for q >.7 a.u., but shows an unexpected relative maximum at q=0, which is proper of s-type momentum densities. Moreover, the intensity is only 20% of the H.F. value. Although presence of a 5s component can not be ruled out, due to the limited experimental energy resolution, simple overposition of $4d^{-1}$ transition and $5s^{-1}$ satellites does not explain the momentum profile measured in coplanar symmetric kinematics. This behaviour can not be explained within C.I. scheme. Amusia (29) has proposed a "melting" of the orbital concept to explain similar evidences observed in photoelectron spectroscopy.

Strong correlation effects bring to complete breakdown of the single particle model also in the case of intermediate orbitals of

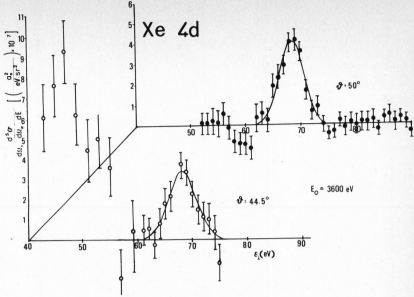

fig. 3 The (e,2e) energy separation spectra of Xe (Ref.28).
Non coplanar symmetric geometry.

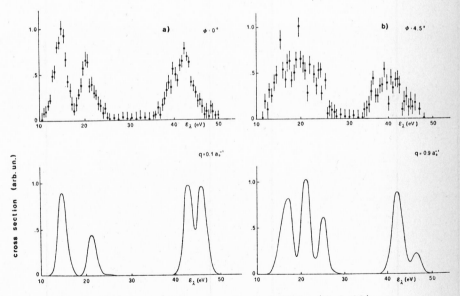

fig. 4 The (e,2e) separation spectra of CHF_3 (Ref.32) at
$q=0.1\ a_o^{-1}$ (a), and $q=0.9\ a_o^{-1}$ (b). Top: experimental
data; bottom: 2ph-TDA calculation.

simple molecules. Cederbaum and Domcke (30) have treated this problem introducing the one-electron Green's function approach. Cambi et al. (31) have applied this method to (e,2e) spectra of the fluorinated series of methane and monoaloderivatives of the C_2H_4 molecule measured by the Frascati group. The experimental separation spectra of CHF_3 is reported in fig. 4 together with predictions done on the basis of a non-diagonal Green's function calculation (32). Complete breakdown of the 2e and $3a_1$ intermediate orbitals is predicted. For these orbitals the single particle model predicts two sharp transitions at 45.3 and 48.2 eV. The Green's function approach shifts the position of the main $2e^{-1}$ transition, spreading its intensity over a large number of satellite peaks. The $(3a_1)^{-1}$ transition splits in two intense peaks plus many other of lower intensity. All of these features are confirmed by the (e,2e) experiment.

CONCLUSION

Impulsive (e,2e) experiments have been reviewed with the aim of pointing out validity of the simple first-order interaction mechanisms for ionization and relevance of electron-electron correlations to the description of the electron structure of atoms and molecules. The following can be concluded:

i) A large momentum transfer is of prime relevance to reliability of simple first order interaction theories (i.e. PWIA) even in the limit of high incident energies. At intermediate energies and symmetric final conditions, simple corrections to the first order interaction model can be done in order to partially account for three-body interaction effects (EWIA). At low and intermediate energies, and anyhow in asymmetric conditions, "corrected" first order model do not account for experiments.

ii) The high energy symmetric (e,2e) events, for which the scattering process is essentially a two-body one, bring information on the electron correlations present in the unperturbed Hamiltonian, which play their role in determining the electron structure of the target. Descriptions beyond single particle scheme are being tried either by in creasingly larger C.I. basis or by Green's function approach. Neither one of them has established a clear superiority jet.

REFERENCES

1. A. Giardini-Guidoni, R. Fantoni, R. Camilloni and G. Stefani, Comm. Atom. Mol. Phys. 10, 107 (1981)
2. E. Weigold and I.E.McCarthy, Adv. At. Mol. Phys. 14, 127 (1978)
3. M. Inokuti, Rev. Mod. Phys. 43, 297 (1971)
4. A. Huetz and J. Mazeau, private communication (1983)
5. E. Schüle and W. Nakel, J. Phys. B15, L639 (1982)
6. E. Weigold, C.J. Noble, S.T. Hood and I. Fuss, J. of Phys. B12, 291 (1979)
7. R. Cambi, G. Ciullo, A. Sgamellotti, F. Tarantelli, R. Fantoni, A. Giardini-Guidoni, I.E. McCarthy and V. Di Martino, Chem. Phys. Lett. 101, 477 (1983)
8. G. Stefani, R. Camilloni and A. Giardini-Guidoni, Phys. Lett. 64A, 364 (1978)
9. A. Giardini-Guidoni, R. Camilloni and G. Stefani, in: "Coherence and correlation in atomic collisions", H. Kleinpoppen and J.F. Williams ed., Plenum, New York (1979)
10. B. van Wingerden, J.T.N. Kimman, M. van Tilburg and F.J. de Heer, J. Phys. B14, 2475 (1981)
11. A. Giardini-Guidoni, R. Fantoni, R. Tiribelli, R. Marconero, R. Camilloni and G. Stefani, Phys.Lett. A77 19 (1980) and references therein quoted
12. I.E.McCarthy and E. Weigold, Phys. Rep. 27C 277 (1976)
13. I.E.McCarthy, this boook (1983)
14. I. Fuss, I.E. McCarthy, C.J. Noble and E. Weigold, Phys. Rev.A17: 604 (1978)
15. A. Giardini-Guidoni, R. Fantoni, V. Di Martino and R. Tiribelli, this book (1983)
16. A. Lahman-Bennani, this book (1983) and references therein quoted
17. R. Camilloni, A. Giardini-Guidoni, I.E. McCarthy and G. Stefani, J. Phys. B13, 397 (1980)
18. R. Camilloni and G. Stefani, Abstr. XII ICPEAC, Gatlinburg, USA p 257 (1981)
19. J.T. Kimman, Pan Guang-Yan, C.W. McCurdy and F.J. de Heer, J. Phys. B: At. Mol. Phys. 22, 4203 (1983)
20. Yu. Popov, I. Bang and J.J. Benayoun, J. Phys. B14,4673 (1981)
    Yu. Popov and J.J. Benayoun, J. Phys. B14, 3515 (1981)
21. Yu. Popov private communication (1982)
22. K.T. Leung and C.E. Brion, Chem. Phys. (1983)
23. L. Frost and E. Weigold, J. Phys. B15, 2531 (1982)
24. J. Mitroy and I. Fuss, J. Phys. B15, L367 (1982)
25. R. Camilloni, G. Stefani, M.A. Coplan, N. Goldstein and J.H.

Moore, Europhysics Conference Abstract 5A, 747 (1981)

26. I. Fuss, R. Glass, I.E. McCarthy, A. Michinton and E. Weigold, J. Phys. B14, 3277 (1981)
27. L. Frost, J. Mitroy and E. Weigold, J. Phys. B16, 203 (1983)
28. A. Giardini-Guidoni, R. Fantoni, R. Marconero, R. Camilloni and G. Stefani, Contributed Paper, XI ICPEAC Kioto (1979), p. 212
29. M. Ya Amusia, Comm. At. Mol. Phys. 10, 179 (1981)
30. L.S. Cederbaum and W. Domcke, Adv. Chem. Phys. 36, 205 (1977)
31. R. Cambi, G. Ciullo, A. Sgamellotti, F. Tarantelli, R. Fantoni, A. Giardini-Guidoni, Chem. Phys. Lett. 80, 295 (1981)
32. R. Cambi, G. Ciullo, A. Sgamellotti, F. Tarantelli, R. Fantoni, A. Giardini-Guidoni, M. Rosi and R. Tiribelli, Chem. Phys. Lett. 90, 445 (1982)

* Work partially supported by an Italy-France scientific cooperative programme.

MULTIPLE SCATTERING APPROACH TO
THE INTERPRETATION OF (e,2e) EXPERIMENTS

C.J. Joachain

Physique Théorique, Faculté des Sciences,

Université Libre de Bruxelles, Belgium

and

Institut de Physique Corpusculaire

Université de Louvain, Louvain-la-Neuve, Belgium

1. INTRODUCTION

The ionization of atoms by electron impact is certainly one of the most interesting processes in the field of atomic collisions. From the point of view of the basic understanding of collision dynamics, it is extremely challenging, since it exhibits the difficulties of several particle problems, coupled with particularly delicate features due to the infinite range of the Coulomb interaction. From the practical point of view, it is a process of great interest for other fields such as plasma physics and astrophysics.

The most detailed information available about the dynamics of single ioniza-tion reactions has been obtained by analyzing triple differential cross sections (TDCS). The TDCS is a measure of the probability that in a single ionization process an incident electron of energy E_0 and momentum k_0 will produce on collision with the target two electrons having energies E_A and E_B and momenta k_A and k_B, emitted respec-tively in the directions (θ_A, ϕ_A) and (θ_B, ϕ_B). Measurements of TDCS in (e,2e) coinci-dence experiments provide a very sensitive probe of the theory of electron-impact ionization, as well as detailed information on the electronic structure of the target.

In this article I shall give a survey of the theory of (e,2e) reactions, with particular emphasis on recent work going beyond first-order calculations. In Section 2 I shall briefly describe several kinematical arrangements and their implications on the theoretical analysis of the collision. Section 3 will be devoted to a theoretical analysis of coplanar asymmetric (Ehrhardt-type) experiments using the eikonal-Born series method. Finally, in Section 4, the theory of coplanar symmetric (e,2e) reactions will be discussed, in both the (e,2e) spectroscopy region and in the large angle region. Atomic units (u.a.) will be used.

2. (e,2e) COINCIDENCE EXPERIMENTS

Let us consider an (e,2e) coincidence experiment in which a fast electron of energy E_0 and momentum k_0 is incident on a target atom. We write the momentum conservation law as $k_0 = k_A + k_B + Q$, where Q is the recoil momentum of the ion. The energy conservation law reads $E_0 - (w_f^{ion} - w_i) = E_A + E_B + T_{ion}$, where w_i is the internal energy of the target atom in the initial state, w_f^{ion} the internal energy of the ion in the final state, and $T_{ion} = Q^2/2M$ the (small) kinetic energy of the recoiling ion of mass M. We denote by $K = k_0 - k_A$ the momentum transfer, or more precisely the momentum lost by the fast or "scattered" electron A, with $k_A \geqslant k_B$. The slow electron B is usually called the "ejected" electron.

It is useful to distinguish between various types of (e,2e) coincidence experiments, depending on the choice of the kinematics. In coplanar geometries the momenta k_0, k_A and k_B are in the same plane, while in non-coplanar geometries the vector k_B is out of the (k_0, k_A) reference plane. Another important distinction can be made between asymmetric and symmetric geometries. In asymmetric geometries, for a given (high) incident electron energy E_0, a fast electron A is detected in coincidence with a slow electron B. The most extensive series of experiments of this kind have been performed since 1969 by Ehrhardt et al. [1-5] in helium and other noble gases. They selected a coplanar geometry and chose the scattering angle θ_A of the fast electron to be fixed and small, while the angle θ_B of the slow electron was varied [see Fig. 1(a)]. It is important to note that in this Ehrhardt-type asymmetric geometry the magnitude K of the momentum transfer is small. Recently, Lahmam-Bennani et al. [6,7] have also measured TDCS in noble gases using a coplanar asymmetric geometry, with incident electrons having high energies (about 8 keV). Ehrhardt-type coplanar asymmetric (e,2e) experiments carried out in atomic hydrogen by Lohmann and Weigold [8,9] were also reported during this Symposium.

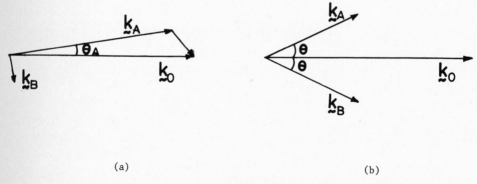

(a)   (b)

Fig. 1. Schematic diagrams of the kinematics of (a) a coplanar asymmetric and (b) a coplanar symmetric (e,2e) reaction.

Symmetric geometries are such that $\theta_A \simeq \theta_B$ and $E_A \simeq E_B$. The first (e,2e) symmetric experiments were performed by Amaldi et al. [10] and since then several series of measurements using coplanar or non-coplanar symmetric geometries have been made, in particular by Camilloni et al. [11,12], Weigold et al. [13-15], de Heer et al. [16,17] and Pochat et al. [18-20]. As an example, let us consider a coplanar symmetric geometry with $E_A = E_B$, $\theta_A = \theta_B(=\theta)$, $\phi_A = 0$, $\phi_B = \pi$ [see Fig. 1(b)] and E_0 large. Then $k_A = k_B \simeq k_0/\sqrt{2}$, $K \simeq k_0(3/2-\sqrt{2} \cos \theta)^{1/2}$ and $Q \simeq k_0|\sqrt{2} \cos \theta - 1|$. We remark that K is never small in this geometry, in contrast with the Ehrhardt-type asymmetric kinematics. For scattering angles $\theta \lesssim 70°$ the values of Q remain small or moderate (with $Q \simeq 0$ when $\theta = 45°$); this angular domain may be called the (e,2e) spectroscopy region since in this case the collision may be considered to be of the impulse type and the momentum distribution of the struck electron can be obtained [10-15]. For $\theta \gtrsim 70°$ both K and Q are large; this region has been explored very recently in the case of (e,2e) reactions in helium [20].

To conclude this section, I would like to mention other interesting (e,2e) experiments which will not be discussed in this article. These are the non-coplanar measurements of Beaty et al. [21,22] in helium and the atomic hydrogen experiments of Weigold et al. [23,24] performed in a kinematical regime intermediate between the asymmetric and symmetric geometries considered here.

3. THEORY OF COPLANAR ASYMMETRIC (e,2e) REACTIONS

In performing their coplanar asymmetric (e,2e) experiments in helium, Ehrhardt et al. [1-5] found a double peaked, strong angular correlation between the scattered and ejected electrons. The first peak near the direction of the momentum transfer $\underset{\sim}{K}$ is called the binary peak; the second peak, near the opposite direction $-\underset{\sim}{K}$ is called the recoil peak. Peaks at precisely the directions $\underset{\sim}{K}$ and $-\underset{\sim}{K}$ are predicted by the first Born approximation. However, Ehrhardt et al. also found the following features which are deviations from the first Born approximation : i) a shift of the binary peak to larger angles, ii) a shift of the recoil peak, also to larger angles and iii) a major enhancement of the magnitude of the recoil peak.

The experimental work carried out since 1969 was steadily accompanied by theoretical treatments of the problem, including first Born calculations [25-27], Coulomb-Born calculations [25,28,29] and distorted-wave Born approximation treatments [30-33], which are all of first order in the electron-electron interaction. Although the Coulomb-Born and distorted-wave treatments are improvements over the first Born approximation, and were able to describe relatively well the binary peak, severe discrepancies remained concerning the size, shape and position of the recoil peak.

The first theoretical treatment in which all the characteristic features of the experiments of Ehrhardt et al. were reproduced was a second Born calculation performed by Byron, Joachain and Piraux [34] in atomic hydrogen, within the framework of the eikonal-Born series (EBS) method [35-37]. According to this method, the direct scattering amplitude is given by

$$f_{EBS} = \overline{f}_{B1} + \overline{f}_{B2} + \overline{f}_{G3} \tag{1}$$

where $\overline{f}_{B1}$ is the first Born amplitude, $\overline{f}_{B2}$ the second Born term and $\overline{f}_{G3}$ the third term of the Glauber series, obtained by expanding the Glauber amplitude in powers of the direct projectile-target interaction V_d. The exchange amplitude g is related to the direct amplitude f by the Peterkop theorem [38], $g(k_A, k_B) = f(k_B, k_A)$, so that we have

$$g_{EBS}(k_A, k_B) = f_{EBS}(k_B, k_A) \tag{2}$$

and the TDCS for an unpolarized e$^-$-H system is given by

$$\frac{d^3\sigma}{d\Omega_A d\Omega_B dE} = \frac{k_A k_B}{k_0} \{ |f_{EBS}|^2 + |g_{EBS}|^2 - Re(f_{EBS}^{\mathbf{x}} g_{EBS}) \} \tag{3}$$

In the case of Ehrhardt-type experiments, for which K is small, the third order Glauber term $\overline{f}_{G3}$ is small. Moreover, exchange effects are also unimportant in this geometry. In addition, the fact that the ejected electron is slow allows one to calculate the second Born term $\overline{f}_{B2}$ approximately by replacing in it the target energy differences $(w_n - w_0)$ by an average excitation energy $\overline{w}$. The sum on intermediate target states can then be done by closure. Calling $\overline{f}_{SB2}$ the second Born term evaluated in this way, the TDCS corresponding to Ehrhardt-type asymmetric geometries is given to good approximation by

$$\frac{d^3\sigma}{d\Omega_A d\Omega_B dE} = \frac{k_A k_B}{k_0} |\overline{f}_{B1} + \overline{f}_{SB2}|^2 \tag{4}$$

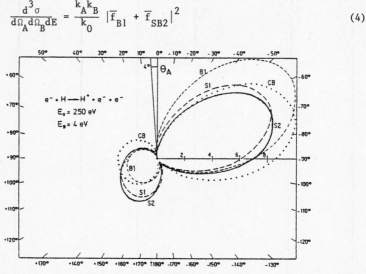

Fig. 2. The TDCS (in a.u.) for the ionization of atomic hydrogen by electron impact, for the case $E_0 = 250$ eV, $E_B = 4$ eV and $\theta_A = 4°$. B1(---) : first Born approximation, S1(-·-) : second Born approximation with $\overline{w} = 1$ a.u.; S2(——) : second Born approximation with $\overline{w} = 0.5$ a.u.; CB(+++) : Coulomb-Born approximation, (from Byron, Joachain and Piraux [34]).

The TDCS for e⁻-H ionization, obtained by Byron, Joachain and Piraux [34] from Eq. (4) is shown in Fig. 2 for the case E_0 = 250 eV, E_B = 4 eV and θ_A = 4°, and for two values $\bar{w}$ = 1 a.u. and $\bar{w}$ = 0.5 a.u. of the average excitation energy. The results corresponding to the first Born and Coulomb-Born approximations are also shown for comparison. It is clear that the second Born results exhibit all the features of the Ehrhardt experiments, in contrast with the first Born and Coulomb-Born calculations.

In view of these encouraging results, the second Born treatment of Eq. (4) was extended by Byron, Joachain and Piraux [39] to the (e,2e) process in helium

$$e^- + He(1^1S) \rightarrow He^+(1s) + e^- + e^- \qquad (5)$$

and a detailed comparison was made with the recent experimental data of Ehrhardt et al. [5]. An additional problem in this case is that of the approximate wave functions used to describe the helium system in the initial and final state. The initial, ground state wave function was chosen to be an analytical fit to the Hartree-Fock wave fonction [40] while the final state was taken to be a symmetrized product of the He⁺(1s) wave function times a Coulomb wave (corresponding to a charge Z = 1), orthogonalized to the ground state Hartree-Fock orbital. It is clear from Figures 3 and 4 that these second Born calculations represent a marked improvement over first order results. In particular, a proper second order treatment of the collision dynamics is essential to account for the angular position of the recoil peak. The remaining discrepancies between theory and experiment, concerning essentially the magnitude of the recoil peak, should probably be taken care of mainly by using more elaborate initial and final state helium wave functions in the second Born calculation.

Recently, a full EBS calculation, using Eq. (3) has been performed by Byron, Joachain and Piraux [41] for the ionization of atomic hydrogen. In addition to the

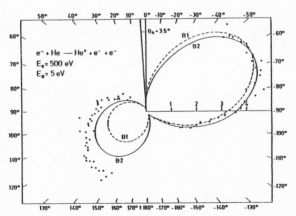

Fig. 3. The TDCS (in a.u.) for the ionization of helium by electron impact, for E_0 = 500 eV, E_B = 5 eV and θ_A = 3.5°. B1(---) : first Born approximation (x 0.87); B2(——) : second Born approximation, calculated with an average excitation energy $\bar{w}$ = 0.9 a.u. The dots are the experimental points. Since the measurements are not absolute, all results are normalized to the same value (at θ_B = - 60°). From Ehrhardt et al. [5].

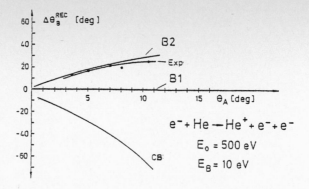

Fig. 4. The angular displacement $\Delta\theta_B^{REC}$ of the recoil peak with respect to the direction $-\underset{\sim}{K}$ ($\underset{\sim}{K}$ being the momentum transfer), as a function of θ_A. B1 : first Born approximation; B2 : second Born approximation of Byron, Joachain and Piraux [39], calculated with $\overline{w} = 0.9$ a.u.; CB : Coulomb-Born approximation. Exp : experimental data of Ehrhardt et al. [5].

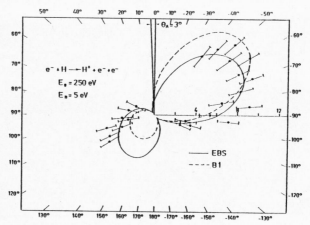

Fig. 5. The TDCS (in a.u.) for the ionization of atomic hydrogen by electron impact, with $E_0 = 250$ eV, $E_B = 5$ eV and $\theta_A = 3°$. B1(---) : first Born approximation; EBS(——) : Eikonal-Born series [41]. The experimental data are the relative measurements of Lohmann and Weigold [8,9]. All results (including a least square fit to the experimental points in the vicinity of the binary peak) are normalized to the same value at $\theta_B = -70°$.

inclusion of the small third order term $\overline{f}_{G3}$ and exchange amplitude g_{EBS}, the calculation of the second order term $\overline{f}_{B2}$ was improved by including exactly the contributions of the 1s, 2s and 2p target states (the other states being included as before by using an average excitation energy $\overline{w}$, which was taken to be 0.5 a.u.). The TDCS obtained in this way for the case $E_0 = 250$ eV, $E_B = 5$ eV and $\theta_A = 3°$ is displayed in Fig. 5, where it is compared with the first Born values and the recent experimental data obtained by Lohmann and Weigold [8,9]. It is clear from Fig. 5 that the measurements are in excellent agreement with the EBS calculation and exhibit all the characteristic features of the helium experiments of Ehrhardt et al. [1-5]. The fact that these features, predicted by the second Born treatment of Byron, Joachain and Piraux [34], are present in the ionization of atomic hydrogen (where both the initial and final states are known exactly) shows unambiguously that they are characteristic of ionization dynamics itself (in particular of second order effects) and not of target correlation effects which arise in many-electron target systems.

4. THEORY OF COPLANAR SYMMETRIC (e,2e) REACTIONS

We have seen in Section 2 that for coplanar symmetric geometries and fast incident electrons one may distinguish two regions depending on the value of the angle θ. For $\theta \lesssim 70°$ the magnitude K of the momentum transfer is relatively large while Q, the magnitude of the recoil momentum of the ion, remains small or moderate. In this case, as well as in non-coplanar symmetric geometries with similar values of K and Q, one has impulse-type, binary encounter collisions, from which the electron momentum density distribution of a variety of target atoms and molecules has been obtained [10-15]. The most elaborate calculations in this (e,2e) spectroscopy region have been performed by using the distorted wave impulse approximation [14,15], in which the transition matrix element has the form

$$T_{ba} = \langle \chi_b | t_M | \chi_a \rangle \tag{6}$$

where χ_a and χ_b are initial and final state distorted waves and t_M is the Mott two-body (electron-electron) Coulomb T-matrix including exchange. It is also interesting to note that in the case of a coplanar symmetric geometry with $\theta = 45°$ (so that $Q \approx 0$) the first Born amplitude $\overline{f}_{B1}$ is of order k_0^{-2} for large k_0 while the second Born term $\overline{f}_{B2}$ is of order k_0^{-3} [42]. The dominance of the first Born approximation for large k_0 in this case is in agreement with the experimental results obtained by Van Wingerden et al. [16,17].

Let us now examine the wide-angle domain ($\theta \gtrsim 70°$) for which both K and Q are large. In this region we expect, on the basis of previous investigations concerning inelastic (excitation) scattering by electron impact [43] that the second Born term $\overline{f}_{B2}$ will be very important, and that this term will be dominated by the contributions of the initial and final target states, acting as intermediate states. This is confirmed by recent calculations performed by Byron, Joachain and Piraux [42,44] for electron impact ionization of atomic hydrogen. They find that for large k_0 and in the case of a coplanar symmetric geometry at large θ the first Born term $\overline{f}_{B1}$ is of order k_0^{-6}, while the contribution of the initial (1s) target state to $\overline{f}_{B2}$ is given by

$$\overline{f}_{B2}(1s) = -\frac{8\sqrt{2}}{\pi} \frac{1}{k_0^5} \frac{1}{(1-\sqrt{2}\cos\theta)^2} \frac{1}{k_0 \cos 2\theta - 2\sqrt{2}\, i \cos\theta} \tag{7}$$

We see from Eq. (7) that $\overline{f}_{B2}(1s)$ is in general of order k_0^{-6}, except at $\theta = 135°$, where it is of order k_0^{-5}. The contribution $\overline{f}_{B2}$ (cont) of the final (continuum) target state to $\overline{f}_{B2}$ may also be estimated for large k_0, and was found to be of order

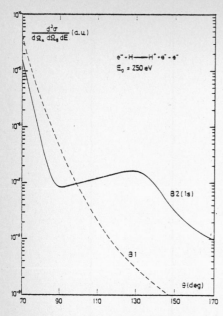

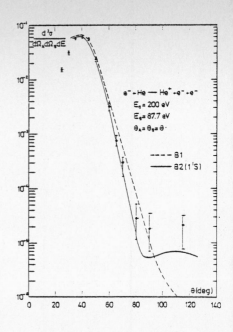

Fig. 6. The TDCS (in a.u.) for the ionization of atomic hydrogen from the ground state by electron impact, for a coplanar symmetric geometry with $E_0 = 250$ eV, $E_A = E_B = 118.2$ eV, as a function of $\theta = \theta_A = \theta_B$. B2(1s) (———) : second Born calculation in which $f = \bar{f}_{B1} + \bar{f}_{B2}(1s)$; B1 (-–-) : first Born approximation.

Fig. 7. The TDCS (in a.u.) for the ionization process $e^- + He(1^1S) \rightarrow He^+ + 2e^-$ for a coplanar symmetric geometry with $E_0 = 200$ eV, $E_A = E_B = 87.7$ eV as a function of $\theta = \theta_A = \theta_B$. B2($1^1S$) (———) : second Born calculation of Byron, Joachain and Piraux, in which $f = \bar{f}_{B1} + \bar{f}_{B2}(1^1S)$ the quantity $\bar{f}_{B2}(1^1S)$ being the helium ground state contribution to $\bar{f}_{B2}$; B1 (-–-) : first Born approximation; ϕ : experimental data of Pochat et al. [20], normalized on theory at $\theta = 40°$.

k_0^{-6} [42]. Thus second order effects due to the term $\bar{f}_{B2}(1s)$ should be particularly important in the large-θ region, expecially near $\theta = 135°$.

This analysis is confirmed by "exact" (numerical) calculations [42,44] of the quantity $\bar{f}_{B2}(1s)$. Fig. 6 shows the TDCS (solid curve) obtained by using $\bar{f}_{B1} + \bar{f}_{B2}(1s)$ as the direct scattering amplitude, and taking into account the fact that the scattering only occurs in the singlet mode. We note the dramatic second order effects in the large angle region, in particular the local minimum around $\theta = 90°$, and the maximum near $\theta = 130°$. This striking behaviour of the TDCS has been observed recently by Pochat et al. [20] in electron-helium ionization experiments, as illustrated in Fig. 7.

REFERENCES

[1] H. Ehrhardt, M. Schulz, T. Tekaat and K. Willmann, Phys. Rev. Letters 22, 89 (1969).

[2] H. Ehrhardt, K.H. Hesselbacher, K. Jung and K. Willmann, in Case Studies in Atomic Physics (North-Holland, Amsterdam) 2, 159 (1971).

[3] D. Paul, K. Jung, E. Schubert and H. Ehrhardt, in The Physics of Electronic and Atomic Collisions (IX ICPEAC, ed. by J.S. Risley and R. Geballe, 1975), p. 194.

[4] H. Ehrhardt, M. Fischer and K. Jung, Zeit. Phys. A 304, 119 (1982).

[5] H. Ehrhardt, M. Fischer, K. Jung, F.W. Byron, Jr., C.J. Joachain and B. Piraux, Phys. Rev. Letters 48, 1807 (1982).

[6] A. Lahmam-Bennani, H.F. Wellenstein, A. Duguet and M. Rouault, J. Phys. B 16, 121 (1983).

[7] A. Lahmam-Bennani, H.F. Wellenstein, C. Dal Cappelo, M. Rouault and A. Duguet, J. Phys. B 16, 2219 (1983).

[8] B. Lohmann and E. Weigold, in Abstracts of papers of the XIII ICPEAC, Berlin (1983), p. 180.

[9] E. Weigold, private communication.

[10] U. Amaldi, Jr., A. Egidi, R. Marconero and G. Pizzella, Rev. Sci. Instr. 40, 1001 (1969).

[11] R. Camilloni, A. Giardini-Guidoni, R. Tiribelli and G. Stefani, Phys. Rev. Letters 29, 618 (1972).

[12] A. Giardini-Guidoni, R. Fantoni, R. Camilloni and G. Stefani, Comments At. Mol. Phys. 10, 107 (1981).

[13] E. Weigold, S.T. Hood and P.J.O. Teubner, Phys. Rev. Letters 30, 475 (1973).

[14] I.E. McCarthy and E. Weigold, Phys. Reports 27, 275 (1976).

[15] E. Weigold and I.E. McCarthy, Adv. At. Mol. Phys. 14, 127 (1978).

[16] B. Van Wingerden, J.T. Kimman, M. Van Tilburg, E. Weigold, C.J. Joachain, B. Piraux and F.J. de Heer, J. Phys. B 12, L 627 (1979).

[17] B. Van Wingerden, J.T. Kimman, M. Van Tilburg and F.J. de Heer, J. Phys. B 14, 2475 (1981).

[18] A. Pochat, R.J. Tweed, M. Doritch and J. Peresse, J. Phys. B 15, 2269 (1982).

[19] A. Pochat, M. Doritch and J. Peresse, Phys. Letters A 90, 354 (1982).

[20] A. Pochat, R.J. Tweed, J. Peresse, C.J. Joachain, B. Piraux and F.W. Byron, Jr., to be published.

[21] E.C. Beaty, K.H. Hesselbacher, S.P. Hong and J.H. Moore, J. Phys. B 10, 611 (1977).

[22] E.C. Beaty, K.H. Hesselbacher, S.P. Hong and J.H. Moore, Phys. Rev. A 17, 1592 (1978).

[23] E. Weigold, S.T. Hood, I. Fuss and A.J. Dixon, J. Phys. B 10, L 623 (1977).

[24] E. Weigold, C.J. Noble, S.T. Hood and I. Fuss, J. Phys. B 12, 291 (1979).

[25] M. Schulz, J. Phys. B 6, 2580 (1973).

[26] V.L. Jacobs, Phys. Rev. A 10, 499 (1974).

[27] W.D. Robb, S.P. Rountree and T. Burnett, Phys. Rev. A 11, 1193 (1975).

[28] S. Geltman and M.B. Hidalgo, J. Phys. B 7, 831 (1974).

[29] S. Geltman, J. Phys. B 7, 1994 (1974).

[30] K.L. Baluja and H.S. Taylor, J. Phys. B 9, 829 (1976).

[31] D.H. Madison, R.V. Calhoun and W.N. Shelton, Phys. Rev. A 16, 552 (1977).

[32] B.H. Bransden, J.J. Smith and K.H. Winters, J. Phys. B 11, 3095 (1978).

[33] R.J. Tweed, J. Phys. B 13, 4467 (1980).

[34] F.W. Byron, Jr., C.J. Joachain and B. Piraux, J. Phys. B 13, L 673 (1980).

[35] F.W. Byron, Jr. and C.J. Joachain, Phys. Rev. A 8, 1267 (1973).

[36] C.J. Joachain, Quantum Collision Theory (North-Holland, Amsterdam, 1975), Chapter 19.

[37] F.W. Byron, Jr. and C.J. Joachain, Phys. Reports 34, 233 (1977).

[38] M.R.H. Rudge, Rev. Mod. Phys. 40, 564 (1968).

[39] F.W. Byron, Jr., C.J. Joachain and B. Piraux, J. Phys. B 15, L 293 (1982).

[40] F.W. Byron, Jr. and C.J. Joachain, Phys. Rev. 146, 1 (1966).

[41] F.W. Byron, Jr., C.J. Joachain and B. Piraux, to be published.

[42] F.W. Byron, Jr., C.J. Joachain and B. Piraux, to be published.
[43] C.J. Joachain, Comments At. Mol. Phys. 10, 107 (1981).
[44] B. Piraux, C.J. Joachain and F.W. Byron, Jr., Abstracts of Papers, XIII ICPEAC (Berlin, 1983), p. 165.

THE ASYMMETRIC (e,2e) REACTION AT HIGH ENERGY

A. Lahmam-Bennani

Laboratoire des Collisions Atomiques et Moléculaires,[†] Université Paris-Sud
Bât. 351, 91405 ORSAY Cedex, FRANCE

To briefly summarize our recent work on the asymmetric high energy (e,2e) reaction, I will use few illustrative examples, limited to the He case, although we have obtained numerous other results on the outer orbitals of Ar and Ne as well as on the 2p orbital of Ar. But first of all, why using high energy and why an asymmetric geometry ?

Roughly speaking, the (e,2e) experiments may be classified into three categories :

- First, symmetrical experiments characterized by large momentum transfer, 3 to 8 a.u. These are the impulsive (or binary) experiments[1] where the description of the collision mechanism is assumed to be known, that is basically the impulse approximation, IA, and the experiment is used to derive information about the target structure. Particularily, the experiment gives a measure of the initial momentum distribution, $\rho(\vec{p})$, of the target electron.

- Non symmetrical or Ehrhardt[2] type experiments which have mainly been performed at low primary energies, 30 to 500 eV, and low momentum transfer values, less than about 1.5 a.u. Here the target structure is assumed to be known, and the experiment is used as a sensitive test for the scattering theory.

- Dipolar experiments[3], with several keV impact energy and small momentum transfer values, about a tenth of an a.u.

The present study is kinematically of the non symmetrical case. However, the high incident energy used (slightly larger than 8 keV) provides a link between these three categories of experiments. Indeed, large variations of the momentum transfer K can be obtained by varying the scattering angle of the fast outgoing electron, θ_a, so that for relatively large values of K and large values of the ejected electron energy, E_b, the IA is valid, allowing a determination of the momentum density $\rho(\vec{p})$. For smaller K values it is possible to study progressive changes in the collision mechanism, from the impulsive limit to the dipolar limit, via the intermediate regime.

This is illustrated in fig. 1 which shows in the usual polar plot our absolute triple differential cross sections, TDCS, for He. These are data taken with 8 keV scattered energy, 100 eV ejected energy, and θ_a values of 1,2,4 and 8°. The main features

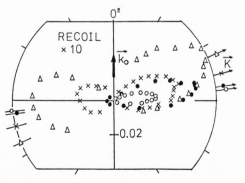

Fig. 1 : Absolute TDCS for He in atomic units for fixed values of θ_a

E_b=100 eV

E_a=8 keV

	θ_a = 1°
×	2°
•	4°
○	8°

here are :

a) the existence of the 2 well-known groups of ejected electrons, forming the binary and the recoil lobes. The binary lobe is peaking in the direction of the momentum transfer $\vec{K}$ (shown by the arrows). The recoil lobe is roughly peaking towards the opposite direction.

b) the rapid decrease of the recoil intensity with increasing scattering angle. At 8°, it becomes zero within the sensitivity of the experiment. This is the impulsive limit where the collision might be considered as a pure binary process. Hence, using the IA framework, the electron momentum distribution, EMD, in the target has been infered from the experiment. This is illustrated in fig. 2 where our experimental absolute $\rho(\vec{p})$ values (with some typical error bars) are compared to different He ground state wavefunctions, a hydrogenic one, a HF - Clementi one and a GTO due to Huzinaga. This demonstrates the ability of the high energy non-symmetrical (e,2e) experiments to measure EMD's and to distinguish between different theoretical wavefunctions.

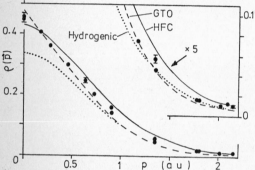

Fig. 2 : Absolute experimental and theoretical momentum density for He

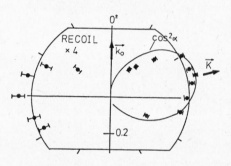

Fig. 3 : Absolute TDCS for He. $E_a=$ 8 keV, $E_b=46$ eV, $\theta_a=1°$, (K=0.4 a.u.) The full curve is the dipolar distribution normalized to the experiment.

c) the drastic reduction in the angular width of the binary lobe, from 65° FWHM at $\theta_a = 1°$ to 20° at 8°.

d) and finaly the observed symmetry of the binary lobe intensity with respect to the $\vec{K}$ direction, while the recoil intensity is not always symmetrical about the $-\vec{K}$ direction. At 4°, the symmetry exists although it is hard to see on the scale used in fig. 1. While at 2 and 1°, the $-\vec{K}$ direction is not anymore a symmetry axis as the lobes seem to be rather centered around 90°. (Similar shifts have already been observed by Ehrhardt and coworkers[2] in their experiments at lower energy, and they were attributed by several authors to the correlation effects). Several arguments can be used to show that, at these small momentum transfer values, the dipolar limit is approached from what concerns the angular position of the lobes as well as their relative intensity or their FWHM. Very briefly, the spectrum of figure 3, taken at 46 eV ejection energy and K=0.4 a.u. shows that a simple $\cos^2$ law, which is the dipolar distribution for He, does quite well in representing the data, but of course only on a relative scale.

After this brief discussion of the two limiting cases, the impulsive one and the dipolar one, let us turn to the comparison between our experiment and theory. Here, a very important point is to be emphasized. All these lobes have more or less the same shape, so that any reasonable theoretical model will reproduce that shape, although these theoretical models might differ by enormous factors, up to 2 or 3 orders of magnitude. It is thus of prime importance to experimentally determine the absolute scale for our data. This is done[4] by comparing the integral of the TDCS over all ejection directions to the corresponding DDCS. This latter one is independently measured and made absolute by making use of the Bethe sum rule[5].

We have essentially considered 3 theoretical models to compare to our experiment. These are : the IA where the target electron ionisation energy is either neglected, or taken into account by modifying the ejected electron wavenumber[4,6] ; a first Born approximation where both outgoing electrons are described by plane waves (OPW) ; or a first Born approximation where the fast scattered electron is represented by a plane wave and the slow ejected one by a Coulomb wave with the effective charge Z as an adjustable parameter (OCW). In both cases the final state of the ion is rewritten orthogonal to the ground initial state of the target.

For the binary lobe, the comparison with experiment is better visualised in the kind of plot shown in figure 4, which represents the intensity at the maximum of the binary lobe as a function of scattering angle θ_a for fixed ejected electron energy, $E_b = 100$ eV and 46 eV . Clearly, the simple IA model works at large ejected electron energies and large momentum transfer values. While the corrected IA (CIA) allows to somehow extend the agreement to the intermediate K values range. They both are one or two orders of magnitude too large at small K values. The 1st Born plane waves model (OPW) does not give any satisfactory agreement with the experiment.

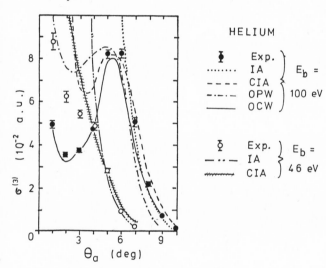

HELIUM

Exp.
IA
CIA $E_b =$
OPW   100 eV
OCW

Exp.
IA $E_b =$
CIA   46 eV

Fig 4 : Absolute TDCS as a function of θ_a. Here $E_a = 8000$ eV, $\theta_b = 80°$ and $82°$ for $E_b = 100$ eV and 46 eV respectively. The 46 eV results have been multiplied by 0.2 in order to be plotted on the same scale.

For $E_b = 100$ eV, the binary intensity is too large at small K values and too small at large K values. This is also true for the recoil intensity, although not shown in figure 4, and is characteristic of this OPW model at all the ejected energies we have investigated, as illustrated on the two examples shown in figure 5, with $E_b = 46$ eV $\theta_a = 2°$, and $E_b = 20$ eV $\theta_a = 1.25°$, respectively. The OPW model, which has been multiplied by 0.7 on this figure, gives a too large intensity for both lobes. Moreover, the example of figure 5b was chosen to emphasize that, although this OPW model gives an intensity which is about a factor of 3 too large for both lobes, it predicts by chance the right ratio for the recoil to binary intensity. This clearly illustrates the importance of comparing the experiment and theory onto an absolute scale.

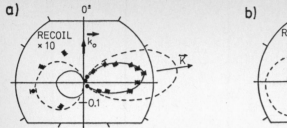 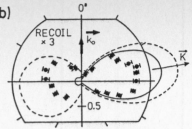

Fig. 5 : Absolute TDCS for He. E_a = 8 keV and a) E_b = 46 eV, θ_a = 2°, b) E_b = 20 eV θ_a = 1.25°. Dashed curve : OPW model multiplied by 0.7, full curve : OCW model.

Finally, good agreement is obtained (fig. 4) with the experiment for the binary intensity calculated using the Born Coulomb wave model (OCW), as shown for the 100 eV case. This is typical of all the spectra we have taken at quite large ejected energies, as also illustrated in figure 5a (E_b = 46 eV) where the binary intensity is correctly reproduced. But this model fails when going to smaller energies (fig. 5b, E_b = 20 eV). And even more severe is the failure of this model for what concerns the recoil intensity, which is always too small, sometimes more than one order of magnitude too small. The simplest way of saying that is to look (fig. 6) at the ratio of the recoil to binary intensity. It is clear on this log-scale that non of the 2 Born approximations, OPW or OCW, does give a satisfactory agreement with the experiment.

Pr. Joachain and coworkers[7] have rencently performed 2nd Born calculations for the Ehrhardt's relative data at 500 eV impact energy. A serious improvement was obtained with respect to the first Born calculations in what concerns this ratio. I think that this 2nd Born model should be applied to data like those of fig. 5, which would also give the opportunity of testing this model onto an absolute scale.

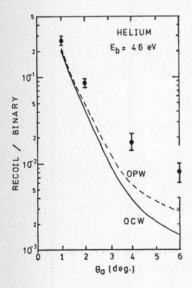

Fig. 6 : Ratio of the recoil to binary intensity. E_a = 8000 eV.

Figure 7 shows the case of argon ionized on its 2p orbital. The situation is far from being as simple as it is for He. All our theoretical models fail completely in reproducing even the shape of the measured lobes.

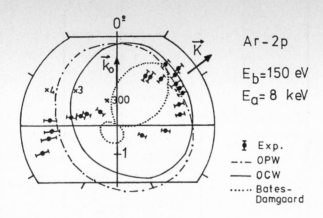

Fig. 7 : Relative TDCS for Ar-2p. θ_a = 2°

[†](laboratoire associé au CNRS)

References

1. G. Stefani, "Impulsive (e,2e) experiments : non dipolar form-factor", this book

2. H. Ehrhardt, K.H. Hesselbacher, K. Jung and K. Willmann J.Phys.B **5**, 1559, 1972

3. C.E. Brion, "Dipolar electron-molecule experiments", this book

4. A. Lahmam-Bennani, H.F. Wellenstein, C. Dal Cappello, M. Rouault and A. Duguet J.Phys.B **16**, 2219, 1983

5. A. Lahmam-Bennani, A. Duguet, H.F. Wellenstein and M. Rouault J.Chem.Phys. **72**, 6398, 1980

6. A. Lahmam-Bennani Phys. Rev. A, accepted for publication, 1983

7. H. Ehrhardt, M. Fischer, K. Jung, F.W. Byron Jr., C.J. Joachain and B. Piraux Phys. Rev. Lett. **48**, 1807, 1982

MULTIPLE SCATTERING APPROACH TO THE (e,2e) REACTION ON ATOMIC HYDROGEN

B. Lohmann, I.E. McCarthy and E. Weigold, The Flinders University of
South Australia, Bedford Park, S.A., 5042, Australia.

It is difficult to formulate the Coulomb three-body problem in a way that leads to a manifestly-convergent approximation scheme for ionization. The coupled-channels-optical theory[1] gives an excellent approximation for scattering, but contains approximate wave functions for ionization channels that are justified only by their ability to reproduce ionization cross sections and the general good quality of the scattering calculations. For explicit ionization reactions we are still at the stage of finding approximations that are justified in various kinematic regions by their ability to reproduce experimental data.

For the electron-hydrogen problem it is convenient to label the incident electron-proton system by 1, the target electron-proton system by 2 and the electron-electron system by 3.

The (e,2e) amplitude is, with obvious kinematic notation,

$$f_{(e,2e)} = \langle \underline{k'}_1 \underline{k}_2' | T | \psi_2 \underline{k}_1 \rangle . \tag{1}$$

The multiple-scattering series for T arises from iterating the three-body equations[2] for finite-range potentials. Up to second order it is

$$T = (t_3 + t_2 G_0 t_3) + (t_1 + t_2 G_0 t_1) + t_1 G_0 t_3 + t_3 G_0 t_1, \tag{2}$$

where t_i is the two-body t-matrix in three-body space for the interaction of the pair i, G_0 is the free-particle Green's function operator. The operators in (2) are grouped to show the relationship to the distorted-wave impulse approximation, which involves some terms up to fourth order.

$$f_{(e,2e)} = \langle \chi^{(-)}(\underline{k'}_1) \chi^{(-)}(\underline{k'}_2) | t_3 + t_1 | \psi_2 \chi^{(+)}(\underline{k}_1) \rangle \tag{3}$$

$$| \chi^{(+)}(\underline{k}_i) \rangle = (1 + G_0 t_i) | \underline{k}_i \rangle . \tag{4}$$

The first term in (3) involves the proton as a spectator in first order. It is expected to apply to the kinematic region involving large momentum transfer to particle 1. After allowing for the identity and spin degeneracy of the electrons the corresponding differential cross section is approximated by

$$\sigma_{(e,2e)} = (2\pi)^4 \frac{k'_1 k'_2}{k_1} f_{ee} | \langle \chi^{(-)}(\underline{k'}_1) \chi^{(-)}(\underline{k'}_2) | \psi_2 \chi^{(+)}(\underline{k}_1) \rangle |^2. \tag{5}$$

This is the factorized distorted-wave impulse approximation, where f_{ee} is the Mott scattering cross section, half-off-shell[3]. The potentials for calculating the distorted waves may be modified to take into account more terms of (2). Such modified potentials are optical potentials, which have been used by Weigold et al.[4] for a 250eV experiment with large momentum transfer. The first-order optical potential for the entrance channel is the ground-state average potential.

Equation (5) is the basis of the approximations to be discussed here. For total energies in the range of some hundred eV, one needs more-and-more detailed descriptions of the distorted waves as the reaction becomes more asymmetric. In the non-coplanar symmetric case[5] relative cross sections are described excellently by using plane waves (fig. 1)

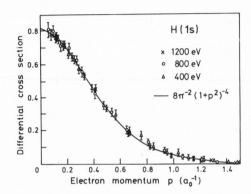

Fig. 1. Non-coplanar distribution of recoil momentum compared with the analytic expression. Experimental points are arbitrarily normalized for each energy.

For this geometry f_{ee} varies very slowly with recoil momentum. The validity of the plane wave approximation is confirmed by the independence of the shape of the momentum profile from the energy. The analytic form for the hydrogen ground state is closely verified.

The main difficulty with the multiple-scattering approach for Coulomb interactions is the three-body boundary condition, a problem related to the logarithmic dependence of the phase of the screened two-body Coulomb t-matrix on the screening parameter[6]. If the three-body wave function is approximated by a product of two distorted waves a necessary condition for choosing effective charges to eliminate the logarithmic phase has been given by Rudge[7]. Weigold et al.[4] showed that the bare charge for both final-state distorted waves gives a good approximation to a wide range of data, while effective charges of the Rudge type give worse agreement. This is illustrated in fig. 2 for a less symmetric experiment at 250eV. The slow electron energy is 50eV. Plane-wave impulse (PWIA) and Born (PWBE) approximations are insufficient for the angular ranges shown. However the distorted-wave impulse approximation works extremely well. DWIA-1 uses the bare Coulomb charges $Z_A = Z_B = -1$. DWIA-4 uses an effective charge -1 for the slow electron and an effective charge consistent with the Rudge constraint[7] for the fast electron.

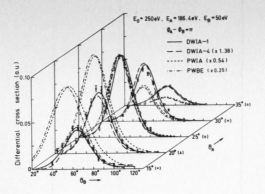

Fig. 2. Differential cross sections for the 250eV (e,2e) reaction on H. The data and calculations are normalized to the peak height of the $\theta_A=25°$ DWIA-1 differential cross section.

For extremely asymmetric conditions the second Born approximation has been used[8]. It is interesting to investigate the low-order structure of various approximations. Neglecting the kinetic energy of the proton the t_1 term of (3) vanishes if $\chi^{(-)}(\underline{k}'_2)$ is calculated with the bare Coulomb potential, since it is then orthogonal to ψ_2. The second order multiple scattering approximation is then identical to the second order distorted-wave impulse approximation.

$$T_{DWIA}^{(2)} = t_3 + t_2 G_0 t_3 + t_1 G_0 t_3 + t_3 G_0 t_1. \tag{6}$$

It is interesting to compare this second-order sum with the corresponding sum from the second Born approximation.

$$T_{2BA} = (1+t_2 G_0)\left[V_1+V_3+(V_1+V_3)(E^{(+)}-K_1-K_2-V_2)^{-1}(V_1+V_3)\right]. \tag{7}$$

Again assuming the bare potential of the nucleus for $\chi^{(-)}(\underline{k}'_2)$ we have to second order in the potentials

$$T_{2BA}^{(2)} = (V_3+V_3 G_0 V_3) + V_2 G_0 V_3 + V_1 G_0 V_3 + V_3 G_0 V_1. \tag{8}$$

This is identical to the second-order potential expansion of (6). The expansions are different in the third order.

The numerical implementation[8] of the second Born approximation (7) involves the closure approximation. This is a further difference from the distorted-wave impulse approximation, whose implementation (5) requires factorization.

Approximation (5), using the ground-state average potential for the entrance channel, is applied to 250eV small-momentum-transfer experiments in figs. 3 and 4. In each case the slow-electron (B) distorted wave is a Coulomb wave with the bare charge. The

arbitrariness inherent in representing the exact final-state wave function by a product of distorted waves is shown in the choice of the fast-electron (A) distorting
potential. We have used the bare charge (dotted curves), plane wave (dot-dash
curves) and the ground-state average potential (full curves). All experimental cross
sections are normalized relative to each other, as are all theoretical cross sections.
The experimental data are normalized to the ground-state-average curve at E_B = 10eV
and θ_A = 5°.

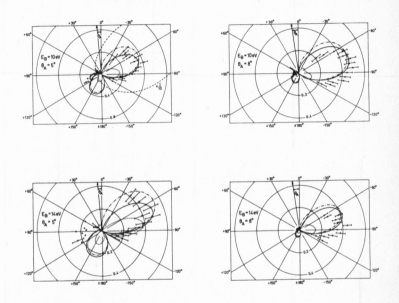

Fig. 3. Polar diagrams for asymmetric differential cross sections for the 250eV
(e,2e) reaction on H. E_B values are 10eV and 14eV. Curves and their
normalization are described in the text.

The bare charge, which works well for fig. 2, gives inferior shapes in all cases.
The ground-state-average approximation, which is intuitively reasonable in view of
the relative speeds of the final-state electrons, works quite well for E_B values
10eV and 15eV (fig. 3), both for direct and recoil peaks. The plane-wave approximation is worse in all cases. None of the approximations are good enough for E_B = 5eV.
The understanding of ionization, which proceeds by an iterative interaction between
experiment and theory, has reached an interesting stage. For symmetric noncoplanar
geometry the simplest form of (5), the PWIA, is sufficient. This approximation is
the basis of the enormously-successful application of the noncoplanar symmetric
(e,2e) reaction to the understanding of wave functions of many-electron atoms and
molecules. Numerical implementations of (5), with particular choices of the fast-
electron potential, are reasonably successful in describing reactions with a high

degree of asymmetry, but not for extreme asymmetry. A slow-electron energy of 5eV for a 250eV experiment is too extreme. It remains for more-fundamental theory to justify the successful choices of fast-electron potentials in the DWIA(5) and to devise successful approximations for extreme asymmetry.

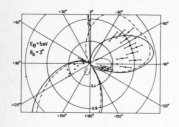

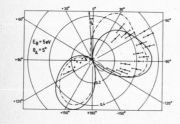

Fig. 4.

Polar diagrams for asymmetric differential cross sections for the 250eV (e,2e) reaction on H. The slow electron energy is 5eV. Curves and their normalization are discussed in the text.

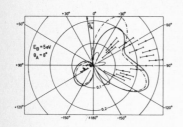

REFERENCES.

1. I.E. McCarthy and A.T. Stelbovics, J. Phys. B 16, 1233 (1983).
2. E. Alt, P. Grassberger and W. Sandhas, Nucl. Phys. B 2, 167 (1967).
3. I.E. McCarthy and E. Weigold, Phys. Rep. 27C, 275 (1976).
4. E. Weigold, C.J. Noble, S.T. Hood and I. Fuss, J. Phys. B 12, 291 (1979).
5. B. Lohmann and E. Weigold, Phys. Lett. 86A, 139 (1981).
6. J.C.Y. Chen and A.C. Chen, Adv. At. Mol. Phys. 8, 71 (1972).
7. M.R.H. Rudge, Rev. Mod. Phys. 40, 564 (1968).
8. F.W. Byron, Jr., C.J. Joachain and B. Piraux, J. Phys. B 15, L293 (1982).

(e,2e) experiments on He in asymmetric conditions
at intermediate energy

V. Di Martino, R. Fantoni, A. Giardini-Guidoni
R. Tiribelli

Comitato Nazionale per la Ricerca e per lo Sviluppo della
Energia Nucleare e delle Energie Alternative (ENEA)
Dip. TIB, Divisione Fisica Applicata, C.R.E. Frascati
C.P. 65 - 00044 Frascati Rome, Italy

Introduction

As it is well known, an (e,2e) reaction consists of an ionization event
occurring on an atomic or molecular target by means of electron impact.
[1] The two outcoming electrons are revealed in coincident time at different energies, satisfying the energy balance $E_0 - \varepsilon_\lambda = E_A + E_B$, where
E_0, E_A and E_B are respectively the incoming and the outcoming electron
energies and ε_λ is the binding energy of the electron ejected from the
target. Momentum conservation equation: $\vec{k}_0 + \vec{q} = \vec{k}_A + \vec{k}_B$ allows to
obtain the momentum distribution $|\psi(\vec{q})|^2$ of the initially bound electron
provided no momentum is retained by the ion.[3] In this reaction the
momentum transfer $\vec{K} = \vec{k}_0 - \vec{k}_A$ is the momentum lost by the scattered
electron. At inciden energy $(E_0 >> \varepsilon_\lambda)$ and high momentum transfer, the
triple differential cross section is well predicted by using plane
(PWIA) or distorted (DWTA) plane waves Impulse Approximation[2].
Furtherly it has been shown[4] that PWIA still holds in asymmetric
conditions at high energy $(E_0 \sim 8$ KeV) provided that $\theta_A \geq \theta_{ee}$, where
$\theta_{ee} = \arcos\sqrt{E_A/E_0}$. On the contrary at low momentum transfer there are
still some discrepancies between the cross section calculated from
different ionization theories and experimental results.[5,6] (e, 2e)
experiments designed to test ionization theories in low momentum
transfer conditions have been performed on noble gases by Ehrarhdt
(1969-1973)[5], later by Beaty (1977-78)[6], and recently by Lahman-
Bennani (1983)[4] and by Pochat.[7] In these experiments the momentum
transfer is lowered by pushing to zero the detection angle of the scattered electron $(\theta_A \to 0)$ and by lowering the energy transferred to the
initially bound electron $(E_0 \simeq E_A >> E_B)$. The He atom whose 1 S orbital
is well described by current theories has been used as target in most
of the cases. s-type electrons give angular distributions characteri-

zed by the presence of two peaks: one, the most relevant called "binary peak" points approximately in the direction of the momentum transfer; the other, pointing at roughly the opposite direction is called "recoil peak". The binary peak can be predicted also by first order[2] theories in the electron-atom interaction potential. At variance many approaches have been tried in order to reproduce well the recoil peak. First satisfactory interpretation of this last peak was given by Geltman[8], who emphasized the role played by the interaction of the slow outcoming electron with the residual ion by using a Coulomb wave to describe this electron. Nevertheless neither the ratio of the two peaks nor their position with respect to the momentum transfer vector was fully reproduced at low incident energy ($E_o \sim$ 250 eV). However it has been recently shown[4] that this approach works quite well at higher energies ($E_o \sim$ 8 KeV) for 0.5 $k\ a_o^{-1}$. Data taken at low E_o are better described by theories including second order terms in the perturbative potential series[9]. Also in this case, however, theoretical results underestimate the recoil peak intensity. This fact gives an indication that the recoil peak is sensitive either to correlation between the two initially bound electrons in the target or to interactions higher than the second order[9] between the fast incoming electron and one of the atomic electrons. These contributions should be reduced as the energy of the incident electron increases while keeping fairly low the ejected electron energy ($10 \leq E_B \leq$ 50 eV) and selecting low momentum transfer ($3° \leq \theta_A$ 15°).

Experimental

General features of this experimental apparatus under computer control are reported in Ref. 3. Its main components are an effusive jet, a non monochromatic commercial electron gun (Varian mod 981-2454) and two emispherical (180°) electrostatic analyzers. These have been supplied with a triple cylindrical element electrostatic lens at the emisphere entrance and a double slit lens at their exit. Angular acceptance is $\Delta\theta_A = \pm 0.2°$ and $\Delta\theta_B = \pm 0.5°$ for each analyzer, which can be independently rotated in the scattering plane in the range $-15° \leq \theta_A \leq +65°$ and $+40° \leq \theta_B + 140°$ with respect to the direction of the incident electrons ($\theta = 0°$). A grid system is placed around the aperture of the lens system of the analyzer lying at small scattering angle (θ_A), with the aim to suppress the incident beam. Under these conditions the momentum transfer resolution obtained is essentially due to the angular divergence of the incident beam $\Delta\theta_o = \pm 1°$ (e.g. at E_o = 2000 eV, $\theta_A = 5°$, E_B = 10 eV, $\Delta k \cong 0.3 a_o^{-1}$). In the measurements here reported an overall energy resolution $1.0 \leq \Delta E_{FWHM} \leq 2.0$ eV is obtained. Helmoltz coils and μ-metal shielding inside the apparatus allowed to reduce the residual magnetic field to less than 3 mgauss in all the scattering region.

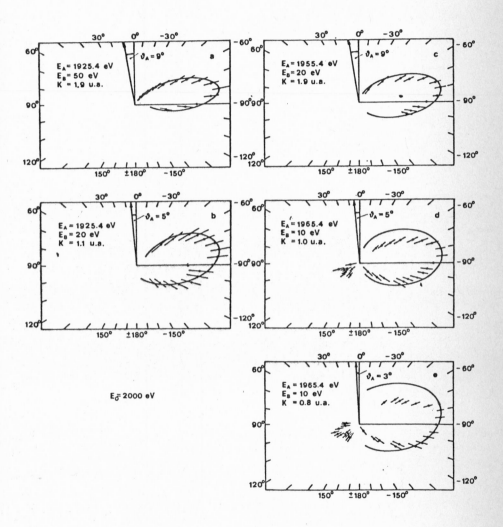

Fig. 1, (e, 2e) spectra in He 1s taken at incident energy $E_o = 2000$ eV compared with PWIA calculations.

Results

Data taken at incident energies E = 2000 eV, 800 eV 500 eV and different momentum transfer are reported respectively in fig. 1, 2, 3
and compared with first order t-matrix calculations. These calculations
have a symmetry axis along the momentum transfer direction. In fig.
1a and 1b the experimental conditions are such that Impulse Approximation holds[2,6] ($E_B > \varepsilon_\lambda$, and no recoil peak is found within experimental errors, as expected. Decreasing E_B or θ_A the higher order
effects are no more negligible and a recoil peak is appearing (fig.1c).
In case of fig. 1d and 1e both θ_A and E_B are under the validity limits
of Impulse Approximation and a large discrepancy with the first order
calculation for the binary peak, as well as a large recoil peak are
present.

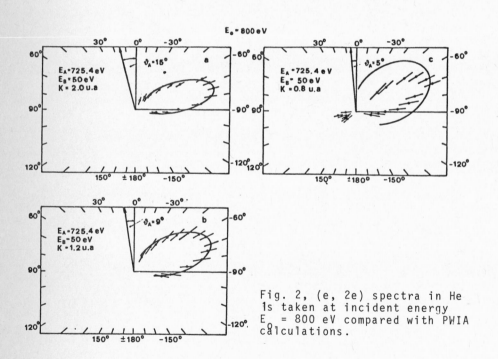

Fig. 2, (e, 2e) spectra in He
1s taken at incident energy
E_0 = 800 eV compared with PWIA
calculations.

An analogous behaviour is observed in data taken at 800 eV and reported in fig. 2. In fig. 2a again a case in which Impulse Approximation
holds is shown, while in fig. 2b the recoil peak starts to appear and
in fig. 2c is clearly present, while the binary peak is shifted towards larger angles and narrowed in the last case.

In fig. 3 are reported data taken at E_0 = 500 eV and E_B and θ_A for which Impulse Approximation does not hold. All the curves show indeed discrepancies with first order predictions even in the shape of the binary peak. A set of data taken at θ_A = 3.5° and E_B = 8 eV is in agreement with previous data[9], as well as data taken at E_0 = 250 eV, θ_A = 10° with E_B = 50 eV and E_B = 20 eV and not reported here for sake of brevity.

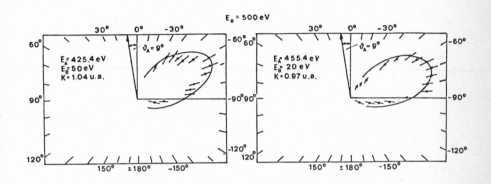

Fig. 3, (e, 2e) spectra in He 1s taken at incident energy E_0=500 eV compared with PWIA calculations.

References

1) I.E. McCarthy, E. Weigold, Phys. Rept., 27 c (1976) 275.

2) R. Camilloni, A. Giardini-Guidoni, I.E. McCarthy, G. Stefani, Phys. Rev. 17 (1978) 1634.

3) R. Fantoni, A. Giardini-Guidoni, R. Tiribelli, R. Cambi, G. Ciullo, A. Sgamellotti, F. Tarantelli, Mol. Phys. 45, (1982) 839.

4) Lahman-Bennani, H.F. Wellenstein, C. Dal Cappello, M. Rouault and A. Duquet, J. Phys. B 16 (1983) 2219.

5) H. Ehrhardt, K.H. Hasselbacher, K. Jung, M. Schultz and Willmann, K., J. Phys. B At. Mol. Phys. 5 (1972), 2107.

6) E.C. Beaty, K.H. Hasselbacker, S.P. Hong, J.H. Moore, J. Phys. B 10 (1977) 611; Phys. Rev. A 17 (1978) 1592.

7) A. Pochat, R.J. Twedd, M. Doritch and J. Peresse, J. Phys. B At. Mol. Phys. 15 (1982) 2269.

8) S. Geltman, M.B. Hidalgo, J. Phys. B 7 (1974) 1994.

9) H. Ehrhardt, M. Fisher, K. Jung, F.W. Byron Jr., C.J. Joachain, and B. Piraux, Phys. Rev. Lett. 48, (1982) 1807.

Coincidence Measurements in (e, 2e) Compton Scattering and Microcanonical Electronic Structures

C. TAVARD, C. DAL CAPPELLO, F. GASSER & M.C. DAL CAPPELLO
Centre "Matières, Rayonnements et Structures, Faculté des Sciences,
Ile du Saulcy - 57045 METZ - FRANCE

Summary

Compton scattering phenomena (X-Ray, Gamma-Ray and electron scattering, positron annihilation...) are commonly admitted to give a direct information on the electronic momentum $\rho(\vec{p})$ of the target.

The Wigner's function $\rho(\vec{r},\vec{p})$ is shown here to give a better physical insight into the description of scattering events. Some preliminary features are brought to support this conclusion.

First-Born and binary encounter assumptions give a unitary formalism applicable to the exact quantum-mechanical description of X-Ray, Gamma-Ray and electron Compton scattering phenomena [1-3] . The second-order differential cross section for the scattered events is shown to be given by

$$d^2\sigma/d\Omega_s \ d(\Delta E) = I_0 <\psi(\vec{r}_1 \ldots \vec{r}_N)| \ \sum_\mu \delta(\Delta E - \frac{k^2}{2} - X + i\vec{k} \ \vec{\nabla}_\mu)|\psi \ (\vec{r}_1 \ldots \vec{r}_N)>$$

$$= (I_0/k) \ J(\vec{k},q). \tag{1}$$

Ω_s defines the scattering angle $(\vec{k}_i,\vec{k}_s)$, $\vec{k} = \vec{k}_i - \vec{k}_s$ represents the momentum transfer and $\Delta E = E_i - E_s$ is the energy lost by the incoming particle. I_0 is a multiplicative factor (depending on either Thomson, Rutherford or Mott scattering cross-sections). I.e., in the specific case of electron scattering and under neglect of exchange :

$$I_o = k_s \ I_{Rutherford} = (2/a_0)^2 \ (k_s/k_i \ k^4). \tag{2}$$

A generalized Compton profile $J(\vec{k},q)$, depending on the usual Compton parameter $q = (\Delta E - k^2/2)/k$, allows to rewrite Eq.(1) as :

$$J(\vec{k},q) = k < \psi \ | \ \sum_\mu \delta(qk - X + i\vec{k}.\vec{\nabla}_\mu)| \ \psi > . \tag{3}$$

Eq. (1) also depends on the electronic structure of the target via the ground-state wave function ψ, which satisfies $(H - E_o) \psi = X \psi = 0$ (Schrödinger equation). Under binary encounter conditions $J(\vec{k},q)$ represents the expectation value of a monoelectronic operator ; for simplicity, we restrict the basic equations to hydrogenic systems.

Impulse assumptions ($[\vec{v}, X] = 0$) for large momentum transfers reduce Eq.(3) to the impulse profile [4] .

$$J_{imp.}(q) = k < \psi | \quad \delta(qk + \vec{k}.\vec{v}) | \psi > =$$

$$= k \int d\vec{p} <\psi | \; p> \delta(qk - \vec{k}\vec{p}) < p|\psi> = \int \rho(\vec{p}) \; \delta(q - p_z)d\vec{p} \tag{4}$$

after uses of the $\chi(\vec{p}) = < p|\psi> = (2\pi)^{-3/2} < \exp(i\vec{p}\vec{r})|\psi>$ momentum representation and the $\rho(\vec{p})$ momentum density.

The above considerations can be extended to the third-order differential cross-section

$$P(\Omega_e) = \frac{d^3\sigma}{d\Omega_s \; d(\Delta E) \; d\Omega_e} = \int \frac{d^4\sigma}{d\Omega_s \; d(\Delta E) \; d\Omega_e \; d(k_e^2/2)} \; d(\frac{k_e^2}{2}) \tag{5}$$

for ionizing processes (E_i ionization energy) in which both scattered particle (momentum $\vec{k}_s$) and ejected electron ($\vec{k}_e$) are observed by coincidence techniques [5,6] on the energy shell ($\Delta E = k_e^2/2 + E_i$) $P(\Omega_e)$ is introduced for convenience to correspond to experimental measurements [5] for fixed Ω_s and ΔE. The following identification

$$\frac{d^2\sigma_{imp.}}{d\Omega_s \; d(\Delta E)} = \int P_{imp.}(\Omega_e) \; d\Omega_e = I_0 \int d\vec{p} \; \rho(\vec{p}) \; \delta(qk - \vec{k}\vec{p}) =$$

$$\simeq \quad I_0 \int d\vec{k}_e \; \rho(\vec{k}_e - \vec{k}) \quad \delta(\Delta E - \frac{k_e^2}{2} - E_i) \tag{6}$$

then gives a probability [7,8] $P_{imp.}(\Omega_e)$ for coincidence events

$$P_{imp.}(\Omega_e) = I_0 \; k_e \; \rho(\vec{k}_e - \vec{k}) \tag{7}$$

This electronic momentum density determination for the corresponding ionized subshell supposes large k momentum transfers, i.e. the following conditions $p_z < p \ll k_e, k$ (or alternatively, to compare the difference $[q - (k_e - k)]$ with respect to k).

Since discrepancies, or Compton defects [9,11], are observed between measurements and accurate $J_{imp.}$ calculations for the scattering cross-section, we may suspect Eq.(7) to be inadequate in most instances. Corrections must be performed [12] for the deflexion of the ejected electron in the electrostatic field of the remaining ion.

With help of closure properties, we then rewrite Eq.(3)

$$J(\vec{k},q) = k \iint d\vec{p}\ d\vec{\lambda}\ <\psi|\ \ \vec{p} - \frac{\vec{\lambda}}{2}>\ \ <\ \vec{p} - \frac{\vec{\lambda}}{2}|\ \delta(qk-X+i\vec{k}.\vec{\nabla})|\ \vec{p} + \frac{\vec{\lambda}}{2}>\ <\vec{p}+\frac{\vec{\lambda}}{2}|\psi>\ . \quad (8)$$

The remaining difficulty arises form the central term. A time-dependent expansion [1,13] can be used to show that exp $[it(X - i\vec{k}.\vec{\nabla})]$ *primarily* contains electrostatic field effects such as $[i\vec{k}\vec{\nabla}, X] = [i\vec{k}\vec{\nabla}, V]$. They commute with $\vec{r}$ coordinates and hence represent multiplicative F operators

$$k\ <\vec{p}- \frac{\vec{\lambda}}{2}\ |\delta(qk-X+i\vec{k}.\vec{\nabla})|\ \vec{p} + \frac{\vec{\lambda}}{2}>\ \simeq (2\pi)^{-3} \int d\vec{r}\ exp(i\vec{\lambda}\vec{r})\ F(\vec{r},\vec{p},\vec{k},q). \quad (9)$$

Under this assumption, we find

$$J(\vec{k},q) \simeq (2\pi)^{-3} \iint d\vec{p}\ d\vec{r}\ F(\vec{r},\vec{p},\vec{k},q) \int d\vec{\lambda}\ exp(i\vec{\lambda}\vec{r})\ <\psi\ |\ \vec{p} - \frac{\vec{\lambda}}{2}>\ <\vec{p} + \frac{\vec{\lambda}}{2}\ |\psi>$$

$$= \iint d\vec{p}\ d\vec{r}\ F(\vec{r},\vec{p},\vec{k},q)\ \ \rho(\vec{r},\vec{p})\ , \quad (10)$$

in which we represent the Wigner's function [14,15] under the peculiar form

$$\rho(\vec{r},\vec{p}) = (2\pi)^{-3} \int d\vec{\lambda}\ \ exp\ (i\vec{\lambda}\vec{r})\ \ \chi^{*}(\vec{p} - \vec{\lambda}/2)\ \ \chi(\vec{p} + \vec{\lambda}/2). \quad (11)$$

As a result of the classical equations of motion for ejected electrons and after use of the $\vec{k}_e$ implicit dependence in $\vec{r}$ and $\vec{p}$ initial conditions, the F function appears somehow connected to an energy conservation, which will simply be represented here by $\delta(\Delta E - k_e^2/2 - E_i)$.

Those arguments given for Eq.(6) then allow, for fixed k_e

$$\int P(\Omega_e)\ d\Omega_e \simeq I_0 \iint d\vec{r}\ d\vec{p}\ \ \rho(\vec{r},\vec{p})\ \ \delta(\Delta E - \frac{k_e^2}{2} - E_i)$$

$$\simeq (I_0/k) \iint d\vec{r}\ \ d\vec{p}\ \ \rho(\vec{r},\vec{p})\ \ \delta(q - p_z)\ . \quad (12)$$

Evaluations of $P(\Omega_e)$ then require a knowledge of $\vec{k}_e(\vec{r},\vec{p},\vec{k})$.
Such a calculation is feasible for the case of a Coulomb field and makes use
of the following construction [16], represented here for an attractive field
(Fig.I). I.e., given $\vec{r} = r\,\vec{u}$ and $\vec{k}'_e = \vec{k} + \vec{p}$ determine the asymptotic momentum
$\vec{k}_e$ ($k_e^2 = k'^2_e - 2Z/r$).

The impulse result (Eq.7) corresponds to $\vec{k}_e(\vec{p},\vec{k})$, independent of $\vec{r}$.
Another extreme situation $\vec{k}_e(\vec{r},\vec{k})$ has been discussed elsewhere [16] and exhibits
a $P(\Omega_e)$ behavior due to the Jacobian (Fig.II)

$$\frac{d\Omega_r}{d\Omega_e} = \frac{(k^2 - k_e^2)^2}{|\vec{k} - \vec{k}_e|^4} \ . \tag{13}$$

For smaller momentum transfers, Eq.(12) predicts a $P(\Omega_e)$ dependence
on both initial $\vec{r}$ and $\vec{p}$. When classical $\rho(\vec{r},\vec{p})$ phase-space models are introduced
in place of the quantum-mechanical Wigner's function, assumed angular correlations
between $\vec{r}$ and $\vec{p}$ may justify the typical aspects of coincidence $P(\Omega_e)$ measurements.
For example, observations of an extra lobe in a direction opposite to $\vec{k}$ can be
explained via hypothesis of a strong angular correlation between $\vec{r}$ and $\vec{p}$ electron
coordinates ($\vec{r}$ and $\vec{p}$ parallel or antiparallel, as shown in Fig. III with two
possible ejection trajectories). Equivalent representations of this planar
configuration are given in Fig. IV and V for a further calculation of $d\Omega_r/d\Omega_e$.

The (e,2e) Orsay's unit [17] allows a number of special conditions
(large $\vec{k}_i$ and $\vec{k}_s$, neglect of exchange, absolute normalization...). Together
with a first-Born binary-encounter Coulomb-wave calculation for $P(\Omega_e)$, recent
measurements [17] for the Helium atom have been used to support our *preliminary*
conclusions about a number of $\rho(\vec{r},\vec{p})$ models. Details of the calculations are too
lengthy to be reproduced here. However, for the very sensitive condition of large
$q = p_z$ values ($q > 5$, i.e. near the nucleus) the best agreement is found for a
model of "electronic breathing", with a strong angular correlation between $\vec{r}$ and $\vec{p}$
coordinates.

We are grateful to A. LAHMAM-BENNANI (Orsay) and H.F. WELLENSTEIN
(Brandeis) for prior communication of their (e, 2e) measurements and to
M.J. COOPER (Warwick) for a number of helpful references. Thanks are due to
Mr P. SENOT and Mrs E. MEDUCIN for technical assistance to this work.

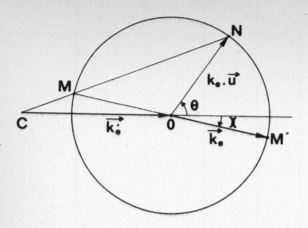

Fig. I - Trajectory deflexion in an attractive Coulomb field. Initial conditions $\vec{r} = r\,\vec{u}$, $\vec{k}'_e = \vec{k} + \vec{p}$ determines the asymptotic momentum $\vec{k}_e$. χ defines the angular deflexion.

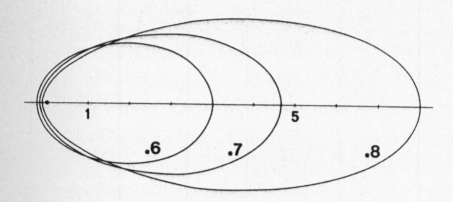

Fig.II - Ehrardt's representation of Eq. (13), for $\mu = .6, .7$ and $.8$
($\mu = 2k \cdot k_e /(k^2 + k_e^2)$).
Lobes point along $\vec{k}$ direction.

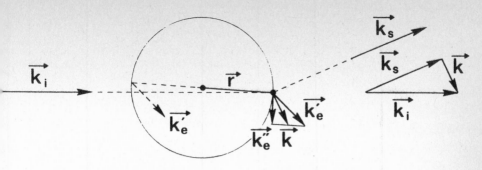

Fig.III – Accounting for $\vec{r}$ and $\vec{p}$ initial conditions $\vec{k}'_e = \vec{k} + \vec{p}$, $\vec{k}''_e = \vec{k} - \vec{p}$ in a classical situation ($\vec{r}$ and $\vec{p}$ here parallel).

Fig. IV and V. – Deflexion of both corresponding trajectories.
$\vec{C_1 0} = \vec{k}'_e = \vec{C0}$, $\vec{C_2 0} = \vec{k}''_e$,
$\overline{HD} = p_z = q$.

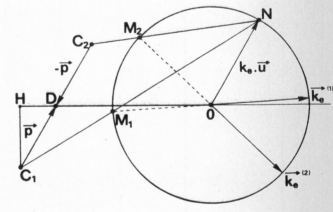

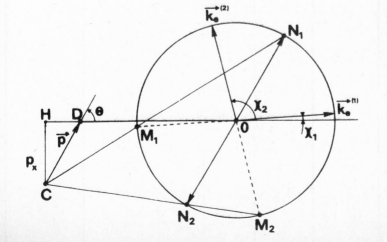

The angular deflexion (χ_1 or χ_2) is defined with respect to $\vec{k} = \vec{D0}$.

1. P. EISENBERGER & P.M. PLATZMAN, Phys. Rev. A 2, 415 (1970).

2. R.A. BONHAM & C. TAVARD, C.R. Acad. Sc. Paris, serie B, 266, 1469 (1968) and 267, 91 (1968).

3. C. TAVARD & R.A. BONHAM, J. Chem. Phys., 50, 1736 (1969).

4. J.W.M. DU MOND, Phys. Rev., 33, 643 (1929).

5. H. EHRARDT, M.S. SCHULZ, T. TEKAAT & K. WILLMANN, Phys. Rev. Lett., 22, 89 (1969).

6. S.T. HOOD, I.E. Mc CARTHY, P.J.O. TEUBNER & E. WEIGOLD, Phys. Rev. A, 8, 2494 (1973).

7. R. CAMILLONI, A. GIARDINI-GUIDONI, I.E. Mc CARTHY & G. STEFANI
    Phys. Rev. A, 17, 1634 (1978).

8. E. WEIGOLD, C.J. NOBLE, S.T. HOOD and I. FUSS, J. Phys. B, Atom. Molec. Phys., 12, 291 (1979).

9. R.J. WEISS, M.J. COOPER & R.S. HOLT, Phil. Mag., 36, 193 (1977).

10. B.J. BLOCH & L.B. MENDELSOHN, Phys. Rev. A, 9, 129 (1974) and 12, 551 (1975).

11. C. TAVARD, M.C. DAL CAPPELLO, F. GASSER, C. DAL CAPPELLO & H.F. WELLENSTEIN, Phys. Rev. A (1982), in press.

12. M.L. GOLDBERGER & K.M. WATSON, Collision Theory, J. Wiley ed. (1964), pp. 730 - 743.

13. F. GASSER & C. TAVARD, Chem. Phys. Lett., 79, 97 (1981).

14. E.P. WIGNER, Phys. Rev. 40, 749 (1932).

15. M.V. BERRY, Phil. Trans. Roy, Soc. London, 287, 237 (1977).

16. C. TAVARD & C. DAL CAPPELLO, C.R. Acad. Sc. Paris, II, 295, 439, (1982)

17. A. LAHMAM-BENNANI, H.F. WELLENSTEIN, C. DAL CAPPELLO, M. ROUAULT & A. DUGUET
    J. Phys B (1983), in press

(e, 3e) collisions : a prospective study

C. DAL CAPPELLO and M.C. DAL CAPPELLO

Centre de Recherches "Matière, Rayonnements et Structures",
Université de METZ, Ile du Saulcy, 57045 METZ CEDEX

Collisions (e, 3e) allow to obtain a detailed information about the electronics correlations. A preliminary calculation concerning a target of Helium is presented here.

Proposed since 1977 by Neudatchin and all [1] the (e, 3e) process consists to detect three electrons in coincidence, resulting from a double ionization of atoms or molecules. The possibility to obtain informations about the electronic correlations of the target is an essential advantage of these experiments. The differential cross-sections (e, 3e) can be shown to be connected with a double-Fourier transform of the initial state wave-function :

$$\Psi_{of}(\vec{K}, \vec{K'}) = (2\pi)^{-3} \int d\vec{r}_1 \ldots d\vec{r}_n \, \Psi_0(\vec{r}_1, \ldots, \vec{r}_n) \, \Psi_f^*(\vec{r}_3, \ldots \vec{r}_n) \, e^{-i\vec{K}.\vec{r}_1} \, e^{-i\vec{K'}.\vec{r}_2}$$

(1)

where $\vec{K}$ and $\vec{K'}$ simultaneously depend of the momentum transfer $\vec{k} = \vec{k}_i - \vec{k}_f$ ($\vec{k}_i$ incident and $\vec{k}_f$ scattered) and the momentum of the two ejected electrons $\vec{k}_e$ and $\vec{k'}_e$. As an example for Helium like systems, the differential cross-section under the first Born approximation is given by :

$$d^6\sigma \Big/ d(k_e^2/2)d(k_e'^2/2) \, d(k_f^2/2) \, d\omega_{k_e} \, d\omega_{k_e'} \, d\omega_{k_f} =$$

$$\delta(k_i^2/2 - k_f^2/2 - k_e^2/2 - k_e'^2/2 - I^{++}) \, M^2 \, \frac{k_f k_e k_e'}{k_i \, k^4}$$

(2)

With $M = \langle \Psi_0(\vec{r}_1, \vec{r}_2) | \, e^{i\vec{k}.\vec{r}_1} + e^{i\vec{k}.\vec{r}_2} | \Psi^\perp \rangle$

(3)

$\Psi^\perp$ represents the wave function of the final state, orthogonalized with the initial state :

$$\Psi^\perp = \Psi - \langle \Psi_0 | \Psi \rangle \, \Psi_0$$

(4)

In order to examine the validity of the "plane waves" approximation, a further calculation consists to represent for the final state, the ejected electrons by two Coulomb waves :

$$\varphi(\vec{k_e}) = \frac{e^{i\vec{k_e}\cdot\vec{r}}}{(2\pi)^{3/2}} \; e^{\pi Z^*/2k_e} \; \Gamma(1 + iZ^*/k_e) \; F(-iZ^*/k_e, 1, -i(k_e r + \vec{k_e}\cdot\vec{r})) \quad (5)$$

Z^* is the effective charge of the nucleus of Helium "seen" by the ejected electrons. The differential croos-section depends thus directly on the expression (1) :

$$\mathbf{M} = \sqrt{2}\left[\Psi_{of}(\vec{k_e} - \vec{k}, \vec{k_e'}) + \Psi_{of}(\vec{k_e}, \vec{k_e'} - \vec{k}) - 2\Psi_{oo}\right] \quad (6)$$

where $\Psi_{oo} = <\varphi_o \mid \varphi(\vec{k_e}) > < \varphi_o \mid \varphi(\vec{k_e'}) > < \varphi_o \mid e^{i\vec{k}\cdot\vec{r}} \mid \varphi_o>$

and represents the wave function of the initial state.

A comparison between the "plane waves" and "Coulomb waves" models lead to :

$k_i^2/2$	$k_e^2/2 = k_e'^2/2$	k	P.W.	C.W.
10 Ke.V	1000 e.V	3.02 a.u	$1.56 \; 10^{-11}$ a.u	$3.31 \; 10^{-14}$ a.u
10 Ke.V	100 e.V	0.61 a.u	$4.74 \; 10^{-4}$ a.u	$2.86 \; 10^{-7}$ a.u
10 Ke.V	50 e.V	0.53 a.u	$1.16 \; 10^{-2}$ a.u	$6.95 \; 10^{-6}$ a.u
10 Ke.V	20 e.V	0.50 a.u	$9.33 \; 10^{-2}$ a.u	$9.70 \; 10^{-5}$ a.u
5 Ke.V	20 e.V	0.40 a.u	0.134 a.u	$1.22 \; 10^{-5}$ a.u

Columns (4) and (5) shows clearly the importance of the wave function describing the ejected electron (the C.W model asymptotically converges towards the P.W model for $Z^* \longrightarrow o$). These differences subsist even for large momentum values of k_e or k_e' (1 ke.V for example).

In order to observe subsequent croos-sections, the importance of the k^{-4} Rutherford factor is also observed.

As a conclusion, the actual orders of magnitude show the present difficulty to realize such experiments [2]. However, a (e, 3e) process involving the ejection of an Auger electron may also be considered to be attractive. Since a detection (in coincidence) of scattered and ejected electrons has been shown to be feasible in (e, 2e) collisions, it would be sufficient to associate a third detector with fixed energy and sufficiently large angular opening.

These experiments would give informations about the electronic correlation existing between electrons belonging to different rings.

|1| Y.F. SMIRNOV, V.G. NEUDATCHIN, A.V. PAVLITCHENKOV and V.G. LEVIN
    Phys. Lett. , 64 A (1977), 31

|2| A. LAHMAM-BENNANI, private communication

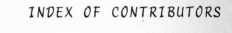

INDEX OF CONTRIBUTORS

INDEX OF CONTRIBUTORS

SUBJECT INDEX

SUBJECT INDEX

Angular-momentum transfer 211
Ar 164,193
ArCl   167
Attachment
    dissociative of DCl   128
                of H2   78,108,123,130
                of HBr   131
                of HCl   126,139
                of HF   122
                of N2   91
                of NaBr   41
    energy of organic molecules 20
Auger   191
Autocorrelation function   157
Autodetachment   14

p-Benzoquinone   14
Bethe-Born   142
    -Goldstone   185
    -ridge   227
BF3   179-181
Binary encounter   227,245,265
    peak   220,241,250,262
Binding energy   156
Born eikonal   242
    first   175,184,192,197,241,251,267
    second   241,252,256

Cd   222,234
Centrifugal barrier 164
    potential   119
Charge density   185
C2H2   27
C2H3Br   226
CH3(NH2)   158
CH4   8,62,86,213
Cl2   167
Close-coupling   199
CO 52,197,207
CO2   35,151,199
Complex potential   119,136
Compton profile   151,208,264
Configuration interaction   176,185,217,233

Continuum functions   98,112
Continuum multiple scattering   86
Correlation effects   206
    electron   151,191,233,271
Coulomb Born   241
    interaction   239,255
    waves   256,267,272
Cr(CO6)   157
Cross-section differential   6,196
    elastic   6,196
    inelastic   6,196
    integral   7,213
    triple differential   192,223,239,250
CS2   151

DCl   128
Delta sigma curve   184
Density matrix   207
Diacetylene   17
Dissociative attachment   122
    states   45,108
Distorted wave   198
    Born   241
    eikonal   229
    impulse   160,229,245,255
Double zeta   152

(e,2e) amplitude   254
    binary   146
    coplanar asymmetric   241-245
    coplanar symmetric   223,245-246
    dipole   145-149o
    noncoplanar symmetric   156
(e,3e)   271
Effective potential   79
Electric-quadrupole transitions   164
Electron trasmission spectroscopy   19
Electronic excitation of H2   53
Electronuclear coupling   132
Energy-loss spectra 14,162,172,210
Equivalent core   167,175
Excitation direct   196
    core   162,172,217

Ion Cyclotron Resonance Spectrometry

Editors: **H. Hartmann, K.-P. Wanczek**

1978. 66 figures, 32 tables. V, 326 pages.
(Lecture Notes in Chemistry, Volume 7)
ISBN 3-540-08760-5

Contents: Line Shapes in Ion Cyclotron Resonance Spectra.
- Quantum Mechanical Description of Collision-Dominated
Ion Cyclotron Resonance. - Improvement of the Electric
Potential in the Ion Cyclotron Resonance Cell. - Thermo-
dynamic Information from Ion-Molecule Equilibrium
Constant Determinations. - Pulsed Ion Cyclotron
Resonance Studies with a One-Region Trapped Ion
Analyzer Cell. - Fourier Transform Ion Cyclotron Reso-
nance Spectroscopy. - Mechanistic Studies of some Gas-
Phase Reactions of $O^{-\cdot}$ Ions with Organic Substrates. -
Studies in the Chemical Ionization of Hydrocarbons. - Gas-
Phase Polar Cycloaddition Reactions. - An Ion Cyclotron
Resonance Study of an Organic Reaction Mechanism. -
Positive and Negative Ionic Reactions at the Carbonyl Bond
in the Gas-Phase. - Ion Chemistry of $(CH_3)_3PCH_2$,
$(CH_3)_3PNH$, $(CH_3)_3PNCH_3$ and $(CH_3)_3PO$.

F. A. Gianturco

The Transfer of Molecular Energies by Collision:

Recent Quantum Treatments

1979. 17 figures, 6 tables. VIII, 327 pages.
(Lecture Notes in Chemistry, Volume 11)
ISBN 3-540-09701-5

Contents: Introduction. - A Résumé of Quantum
Mechanical Potential Scattering. - Potential Energy Hyper-
surface Calculations for Simple Systems. - Rotational and
Vibrational Inelasticity in Molecular Encounters. -
Dimensionality Reduction Methods for Rotovibrational
Cross Section Calculations. - Numerical Methods for the
Coupled Equations: A Survey. - Rotovibrational Relaxation
Models in Simple Gases.

Springer-Verlag
Berlin
Heidelberg
New York
Tokyo

The Unitary Group for the Evaluation of Electronic Energy

Matrix Elements

Editor: **J. Hinze**

1981. VI, 371 pages. (Lecture Notes in Chemistry, Volume 22).
ISBN 3-540-10287-6

Contents: Unitary Group Approach to Many-Electron Correlation Problem. – The Graphical Unitary Group Approach and its Application to Direct Configuration Interaction Calculations. – A Harmonic Level Approach to Unitary Group Methods in CI and Perturbation Theory Calculations. – Many-Body Correlations Using Unitary Groups. – Factorization of the Direct CI Coupling Coefficients into Internal and External Parts. – Multiconfiguration Self-Consistent-Field Wavefunction for Excited States. – Minicomputer Implementation of the Vector Coupling Approach to the Calculation of Unitary Group Generator Matrix Elements. – New Directions for the Loop-Driven Graphical Unitary Group Approach: Analytic Gradients and an MCSCF Procedure. – The Occupation-Branching-Number Representation. – Review of Vector Coupling Methods in the Unitary Group Approach to Many-Electron Problems. – Symmetric Group Graphical Approach to the Configuration Interaction Method. – Orbital Description of Unitary Group Basis. – On the Relation Between the Unitary Group Approach and the Conventional Approaches to the Correlation Problem. – Unitary Bases for X-Ray Photoelectron Spectroscopy. – Broken Unitary Tableaus, Itinerant Nuclear Spins, and Spontaneous Molecular Symmetry Collapse. – CI-Energy Expressions in Terms of the Reduced Density Matrix Elements of a General Reference. – The Unitary Group Formulation of Quantum Chemistry: Generator States. – The Unitary Group Approach to Bonded Functions.

H. Hartmann, K.-P. Wanczek

Ion Cyclotron Resonance Spectrometry II

1982. XV, 538 pages. (Lecture Notes in Chemistry, Volume 31).
ISBN 3-540-11957-4

In ion cyclotron resonance (ICR) spectrometry rapid progress has been made in the past few years. This volume covers recent research in the entire field of ICR in great detail. Almost all the ICR research groups have contributed. There are review papers as well as original contributions and short notes. The articles are arranged in four groups:
- studies of basic properties of molecules and ions (photochemistry, photodissociation, ion energetics, electron affinity)
- ion chemistry (structure and reactivity of positive and negative ions, metal ions)
- instrumentation (Fourier transform ICR spectrometry, tandem instruments, frequency scanning detectors, high magnetic field)
- theory (ion-molecule reactions, ion motion and detection)

Springer-Verlag
Berlin
Heidelberg
New York
Tokyo